Debris Control Structures
Evaluation and Countermeasures
Third Edition

1. Report No. FHWA-IF-04-016 HEC-9	2. Government Accession No.	3. Recipient's Catalog No.		
4. Title and Subtitle Debris Control Structures – Evaluation and Countermeasures Hydraulic Engineering Circular 9 Third Edition		5. Report Date October 2005		
		6. Performing Organization Code HIBT-20, HFL-15		
7. Author(s) J.B. Bradley, D.L. Richards, C.D. Bahner		8. Performing Organization Report No.		
9. Performing Organization Name and Address WEST Consultants, Inc. 2601 25th Street Suite 450 Salem, Oregon 97302		10. Work Unit No. (TRAIS)		
		11. Contract or Grant No. DTFH61-01-Z-00012		
12. Sponsoring Agency Name and Address Office of Bridge Technology FHWA, Room 3203 400 Seventh Street, SW Washington, D.C. 20590 National Highway Institute 4600 North Fairfax Dr., Suite 800 Arlington, Virginia 22203		13. Type of Report and Period Covered Final Report		
		14. Sponsoring Agency Code		
15. Supplementary Notes Project Manager: Brian Beucler, FHWA Technical Assistants: Joe Krolak, Abbi Ginsberg, Jorge E. Pagán-Ortiz, FHWA; Arthur Parola, University of Louisville				
16. This circular provides information on debris accumulation and the various debris control countermeasures available for culvert and bridge structures. It is a dual unit update of the first edition published in 1964. This circular presents various problems associated with debris accumulation at culvert and bridge structures, provides a procedure for estimating the potential of debris accumulating at a bridge structure, and provides general guidelines for analyzing and modeling debris accumulation on a bridge structure to determine the impacts that the debris would have on the water surface profile through the bridge structure and the hydraulic loading on the structure. Various types of debris countermeasures for culvert and bridge structures are discussed within this circular. General criteria for selection of these countermeasures and general design guidelines for some of the structural measures are also included in this manual.				
17. Key Words debris accumulation, debris, culverts, bridges, hydraulic loading, debris-control structures, debris impacts, debris countermeasures, debris management, debris classification		18. Distribution Statement This document is available to the public from The National Technical Information Service Springfield, VA 22151 (703) 487-4650 http://www.fedworld.gov/ntis/		
19. Security Classif. (of this report) Unclassified	20. Security Classif. (of this page) Unclassified		21. No. of Pages 179	22. Price

TABLE OF CONTENTS

(page intentionally left blank)

LIST OF FIGURES

LIST OF TABLES

LIST OF SYMBOLS

A	=	Cross sectional area of flow, m^2 (ft^2)
A_B	=	Net area of the bridge opening, m^2 (ft^2)
A_{bu}	=	Net area of the bridge opening at the upstream face of the bridge, m^2 (ft^2)
A_{bd}	=	Net area of the bridge opening at the downstream face of the bridge, m^2 (ft^2)
A_c	=	Unobstructed cross-sectional flow area in the contracted section, m^2 (ft^2)
A_d	=	Cross-sectional flow area blocked by debris in the contracted bridge section, m^2 (ft^2)
A_D	=	Area of wetted debris based on the upstream water surface elevation projected normal to the flow direction, m^2 (ft^2)
A_{hd}	=	Area of the vertically projected, submerged portion of the debris accumulation below the downstream water surface, m^2 (ft^2)
A_{hu}	=	Area of the vertically projected, submerged portion of the debris accumulation below the upstream water surface, m^2 (ft^2)
A_o, A_i	=	Outlet and inlet storm drain cross-sectional areas, m^2 (ft^2)
A_o	=	Orifice area, m^2 (ft^2)
A_{rack}	=	Area of debris rack, m^2 (ft^2)
B	=	Blockage ratio
C	=	Expansion and contraction loss coefficients
C_g	=	Discharge coefficient for sluice gate type of pressure flow
C_d	=	Discharge coefficient for fully submerged pressure flow
C_D	=	Drag coefficient
C_w	=	Discharge coefficient for weir flow
DB_{EL}	=	Bottom elevation of the debris accumulation, m (ft)
D	=	Culvert diameter, m (ft)
DHW	=	Design high water elevation, m (ft)
D_o	=	Outlet pipe diameter, m (ft)
D_{50}	=	Mean riprap size, m (ft)
E_t	=	Total energy, m (ft)
F	=	Total segment force on the bridge structure, N (lbs)
F_D	=	Drag force, N (lbs)
F_{DEL}	=	Elevation of drag force on the bridge structure, m (ft)
F_{DST}	=	Station of drag force on the bridge structure, m (ft)
F_{EL}	=	Elevation of total segment force on the bridge structure, m (ft)
F_f	=	External force due to friction, N (lbs)
F_h	=	Total hydrostatic force on the bridge structure, N (lbs)
F_{hd}	=	Hydrostatic force on downstream side of the bridge structure, N (lbs)
F_{hEL}	=	Elevation of hydrostatic force on the bridge structure, m (ft)
F_{hu}	=	Hydrostatic force on upstream side of the bridge structure, N (lbs)
F_{hST}	=	Station of hydrostatic force on the bridge structure, m (ft)
Fr	=	Froude number
F_{ST}	=	Station of total force on the bridge structure, m (ft)
g	=	Acceleration due to gravity, 9.81 m/s^2 (32.2 ft/s^2)
h	=	Vertical distance from water surface to center of gravity of flow area, m (ft)

h_f	=	Friction loss, m (ft)
h_v	=	Velocity head, m (ft)
H	=	Increase in water surface elevation from the downstream side to the upstream side of the bridge, m (ft)
H	=	The difference in the upstream energy gradient elevation and the downstream water surface elevation, m (ft)
H	=	Height of debris-control structure, m (ft)
h_{cu}	=	Vertical distance from the upstream water surface to the centroid of area A_{hu}, m (ft)
h_{cd}	=	Vertical distance from the downstream water surface to the centroid of area A_{hd}, m (ft)
HGL_i	=	Hydraulic grade line elevation at the inflow pipe, m (ft)
HGL_o	=	Hydraulic grade line elevation relative to the outlet pipe invert, m (ft)
h_L	=	Energy head loss, m (ft)
H_w	=	Difference between the upstream energy and the road crest, m (ft)
INV	=	Inlet invert elevation, m (ft)
K	=	Conveyance, m^3/s (ft^3/s)
K	=	Yarnell's pier shape coefficient
K_c	=	Units conversion factor or coefficient
K_e	=	Expansion coefficient
L	=	Horizontal length of curve, flow length, length of basin at base length of pipe, or length of culvert, m (ft)
L_w	=	Effective length of the weir, m (ft)
n	=	Manning's roughness coefficient
P	=	Wetted perimeter, m (ft)
P	=	Hydrostatic pressure force, N (lbs)
Q	=	Flow, m^3/s (ft^3/s)
Q	=	Total discharge through the bridge opening, m^3/s (ft^3/s)
Q_w	=	Total discharge over the roadway approaches and the bridge, m^3/s (ft^3/s)
R	=	Hydraulic radius (flow area divided by the wetted perimeter), m (ft)
s	=	Spacing between the bars of a debris-control structure, m (ft)
S	=	Surface slope, m/m (ft/ft)
S_f	=	Friction slope, m/m (ft/ft)
S_L	=	Longitudinal slope, m/m (ft/ft)
S_o	=	Energy grade line slope, m/m (ft/ft)
SL	=	Main channel slope, m/km (ft/mi)
t	=	Bar thickness, m (ft)
T	=	Surface width of open channel flow, m (ft)
V	=	Mean velocity, m/s (ft/s)
V	=	Storage volume, m^3 (ft^3)
V_r	=	Reference velocity, m/s (ft/s)
V_2	=	Mean velocity for the cross-section at the downstream side of the bridge, m/s (ft/s)
V_3	=	Average flow velocity at the cross section immediately upstream of the bridge, m/s (ft/s)
y	=	Flow depth, m (ft)

y_r	=	Average flow depth corresponding with the reference velocity, m (ft)
Y_3	=	Hydraulic depth at the cross section immediately upstream of the bridge, m (ft)
w	=	Width of debris-control structure, m (ft)
W	=	Force due to weight of water in the direction of flow, N (lbs)
W_D	=	Width of debris accumulation defined by design log length, m (ft)
WS_{DS}	=	Water surface elevation downstream of the bridge, m (ft)
W_{min}	=	Minimum width of debris rack, m (ft)
WS_{US}	=	Water surface elevation upstream of the bridge, m (ft)
Z	=	Elevation above a given datum, m (ft)
z	=	Horizontal distance for side slope of trapezoidal channel, m (ft)
γ	=	Specific weight of water, 9810 N/m^3 (62.4 lb/ft^3) at 15.6 EC (60 EF)
γ_s	=	Specific weight of sediment particle, N/m^3 (lb/ft^3)
τ	=	Average shear stress, Pa (lb/ft^2)
ρ	=	Fluid density, kg/m^3 (slugs/ft^3)
ω	=	Ratio of velocity head to depth for the cross-section at the downstream side of the bridge
α	=	Obstructed area of the piers divided by the total unobstructed area for the cross section at the downstream side of the bridge
α	=	Apex angle for a culvert debris deflector, degrees

ACKNOWLEDGMENTS

First Edition

Mr. G. Reihsen, Highway Engineer with the Federal Highway Administration wrote the first edition of this Hydraulic Engineering Circular. The manual was dated February 1964.

Second Edition

Messrs. G. Reihsen and L.J. Harrison, FHWA revised the first edition in March 1971. In that second edition, the authors incorporated comments from practitioners and added additional remarks on safety.

Third Edition

This third edition substantially revises these earlier HEC-9 editions, adding new information, while keeping important portions of the original documents. The authors of this third edition wish to acknowledge the contributions made by the authors of earlier editions.

The writers wish to also acknowledge the technical assistance of Dr. Arthur Parola, Professor at University of Louisville. Finally, the authors wish to thank everyone at FHWA who was involved in the preparation of this manual.

GLOSSARY

abutment:
The structural support at either end of a bridge usually classified as spill-through or vertical.

aggradation:
General and progressive buildup of the longitudinal profile of a channel bed due to sediment deposition.

alluvium:
Unconsolidated material deposited by a stream in a channel, floodplain, alluvial fan, or delta.

average velocity:
Velocity at a given cross section determined by dividing discharge by cross-sectional area.

backwater:
The increase in water surface elevation relative to the elevation occurring under natural channel and floodplain conditions. It is induced by a bridge or other structure that obstructs or constricts the otherwise unobstructed flow of water in a channel.

backwater area:
The low-lying lands adjacent to a stream that may become flooded due to backwater.

bank:
The side slopes of a channel between which the flow is normally confined.

bank, left (right):
The side of a channel as viewed in a downstream direction.

bankfull discharge:
Discharge that, on the average, fills a channel to the point of overflow.

bar:
An elongated deposit of alluvium within a channel, not permanently vegetated.

bed:
The bottom of a channel bounded by banks.

bed load:
Sediment that is transported in a stream by rolling, sliding, or skipping along the bed or very close to it; considered to be within the bed layer.

bed material:
Material found on the bed of a stream (May be transported as bed load or in suspension).

boulder:
A rock whose diameter is greater than 250 mm.

bridge opening:	The cross sectional area beneath a bridge that is available for conveyance of water.
bridge waterway:	The area of a bridge opening available for flow, as measured below a specified stage and normal to the principal direction of flow.
channel:	The bed and banks that confine the surface flow of a stream.
channelization:	Straightening or deepening of a natural channel by artificial cutoffs, grading, flow-control measures, or diversion of flow into a man-made channel.
clay:	A particle whose diameter is in the range of 0.00024 to 0.004 mm.
cobble:	A rock whose diameter is in the range of 64 to 250 mm.
constriction:	A natural or artificial control section, such as a bridge crossing, channel reach or dam, with limited flow capacity in which the upstream water surface elevation is related to discharge.
contraction:	The effect of a natural or man-made channel constriction on flow streamlines.
countermeasure:	A measure intended to prevent, delay or reduce the severity of stream or river problems.
contraction scour:	Contraction scour, in a natural channel or at a bridge crossing, involves the removal of material from the bed and banks across all or most of the channel width. This component of scour results from a contraction of the flow area at the bridge which causes an increase in velocity and shear stress on the bed at the bridge. The contraction can be caused by the bridge or from a natural narrowing of the stream channel.
cross section:	A section normal to the trend of a channel or flow.
culvert:	A drainage conduit that conveys flow from one side of an embankment to the other.
dam jam:	A type of debris jam that extends entirely across the channel as a result of the debris length being approximately equal to the channel width.

debris:	Floating or submerged material, such as logs, vegetation, or trash, transported by a stream.
debris accumulation:	The collection of debris material on a fixed object.
debris cribs:	Open crib-type structures placed vertically over the culvert inlet in log-cabin fashion to prevent inflow of coarse bedload and light floating debris.
debris dams and basins:	Structures placed across well-defined channels to form basins that impede the streamflow and provide storage space for deposits of detritus and debris.
debris deflectors:	Structures placed at the culvert inlet to deflect the major portion of the debris away from the culvert entrance.
debris fins:	Walls built in the stream channel upstream of a culvert or bridge. Their purpose is to align debris, such as logs, with the axis of the culvert or bridge so that the debris will move through the culvert or bridge opening.
debris jam:	Accumulation of floating or neutrally buoyant debris material formed around large, whole trees that may be anchored to the bed or banks at one or both ends, once in the stream system.
debris racks:	Structures placed across the stream channel to collect the debris before it reaches the culvert entrance. Debris racks are usually vertical and at right angles to the streamflow, but they may be skewed with the flow or inclined with the vertical.
debris risers:	A closed-type structure placed directly over the culvert inlet to cause deposition of lowing debris and fine detritus before it reaches the culvert inlet.
deflector jam:	A type of debris jam that redirects the flows to one or both of the banks. These types of jams usually occur when the channel width is slightly greater than the average tree height.
degradation (bed):	A general and progressive (long term) lowering of the channel bed due to erosion over a relatively long channel length.

design log length: A length above which logs are insufficiently abundant and insufficiently strong throughout their full length to produce an accumulation equal to their length. This length does not represent the absolute maximum length of trees within the watershed upstream of the site.

detritus: Non-debris sediment or bed load characterized as *fine* or *course*. Fine detritus is a fairly uniform bed load of silt, sand, gravel more or less devoid of floating debris, tending to deposit upon diminution of velocity. Coarse detritus consists of coarse gravel or rock fragments.

dike: An impermeable or semi-permeable linear structure for the control or containment of overbank flow. A dike-trending parallel with a streambank differs from a levee in that it extends for a much shorter distance along the bank, and it may be surrounded by water during floods.

dike (groin, spur, jetty): A structure extending from a bank into a channel that is designed to: (a) reduce the stream velocity as the current passes through the dike, thus encouraging sediment deposition along the bank (permeable dike); or (b) deflect erosive current away from the streambank (impermeable dike).

drift: Alternative term for "debris" that is floating on or through a river.

eddy current: A vortex-type motion of a fluid flowing contrary to the main current, such as the rotational water movement that occurs when the main flow becomes separated from the bank.

effective length of debris: The length of the debris element that can support the load of the debris accumulation.

ephemeral stream: A stream or reach of stream that does not flow for parts of the year. As used here, the term includes intermittent streams with flow less than perennial.

erosion: Displacement of soil particles on the land surface due to water or wind action.

floodplain: A nearly flat, alluvial lowland bordering a stream, that is subject to frequent inundation by floods.

flow-control structure: A structure either within or outside a channel that acts as a countermeasure by controlling the direction, depth, or velocity of flowing water.

Froude number: A dimensionless number that represents the ratio of inertial to gravitational fluid forces. High Froude numbers can be indicative of high flow velocity and the potential for scour.

geomorphology /morphology: That science that deals with the form of the Earth, the general configuration of its surface, and the changes that take place due to erosion and deposition.

gravel: A rock fragment with a diameter ranging from 2 to 64 mm.

groin: A structure extending from the bank of a stream in a direction transverse to the current. Many names are given to this structure, the most common being "spur," "spur dike," "transverse dike," "jetty," etc. Groins may be permeable, semi-permeable, or impermeable.

guide bank: An embankment extending from the approach embankment at either or both sides of the bridge opening to direct the flow through the opening. Some guide banks extend downstream from the bridge (also see spur dike).

helical flow: Three-dimensional movement of water particles along a spiral path in the general direction of flow. These secondary-type currents are of most significance as flow passes through a bend; their net effect is to remove soil particles from the cut bank and deposit this material on the point bar.

hydraulic problem: An effect of stream flow, tidal flow, or wave action such that the integrity of the highway facility is destroyed, damaged, or endangered.

hydraulic radius: The ratio of a channel's cross sectional area to its wetted perimeter.

ice debris: Accumulation or transport of ice floes in the waterway.

inlet: Entrance of the culvert at the upstream end.

island:	A permanently vegetated area, emergent at normal stage that divides the flow of a stream. Islands originate by establishment of vegetation on a bar, by channel avulsion, or at the junction of a minor tributary with a larger stream.
large floating debris:	Type of debris consisting of trees, logs, and other organic matter with a length greater than 3.5 feet. Also referred to as Large Woody Debris (LWD)
lateral erosion:	Erosion in which the removal of material is progressing primarily in a lateral direction, as contrasted with degradation and scour that progress primarily in a vertical direction.
light floating debris:	Type of debris consisting of small limbs or sticks, orchard prunings, tules, and refuse.
levee:	An embankment, generally landward of top bank, that confines flow during high-water periods, thus preventing flooding into lowlands.
local scour:	Removal of material from around piers, abutments, spurs, and embankments caused by an acceleration of flow and resulting vortices induced by obstructions to the flow.
medium floating debris:	Type of debris consisting of tree limbs, and large sticks.
mid-channel bar:	A bar lacking permanent vegetal cover that divides the flow in a channel at normal stage.
migration:	Change in position of a channel by lateral erosion of one bank and simultaneous accretion of the opposite bank.
outlet	The downstream end of a culvert.
overbank flow:	Water movement that overtops the bank either due to stream stage or to overland surface water runoff.
parallel jam:	A type of debris jam that is oriented parallel to the flow. These types of jams usually occur when the channel width is significantly greater than the maximum debris length.

perennial stream: A stream or reach of a stream that flows continuously for all or most of the year.

pressure flow scour: The increase in local scour at a pier subjected to pressure (or orifice) flow as a result of flow being directed downward towards the bed by the superstructure (vertical contraction of the flow) and by increasing the intensity of the horseshoe vortex. The vertical contraction of the flow can be the more significant cause of the increased scour depth.

reach: A segment of stream length that is arbitrarily bounded for purposes of study.

revetment: Rigid or flexible armor placed to inhibit scour and lateral erosion (see bank revetment).

riprap: In the restricted sense, layer or facing of rock or concrete dumped or placed to protect a structure or embankment from erosion; also the broken rock or concrete suitable for such use. Riprap has also been applied to almost all kinds of armor, including wire-enclosed riprap, grouted riprap, sacked concrete, and concrete slabs.

river training: Engineering works with or without the construction of embankment, built along a stream or reach of stream to direct or to lead the flow into a prescribed channel. Also, any structure configuration constructed in a stream or placed on, adjacent to, or in the vicinity of a streambank that is intended to deflect currents, induce sediment deposition, induce scour, or in some other way alter the flow and sediment regimes of the stream.

roughness coefficient: Numerical measure of the frictional resistance to flow in a channel, as in the Manning's or Chezy's formulas.

sand: A rock fragment whose diameter is in the range of 0.062 to 2.0 mm.

scour: Erosion of streambed or bank material due to flowing water; often considered as being localized (see local scour, contraction scour, total scour).

scoured depth: Total depth of the water from water surface to a scoured bed level (compare with "depth of scour").

sediment: Fragmental material transported, suspended, or deposited by water.

sediment yield:	The total sediment outflow from a watershed or a drainage area at a point of reference and in a specified time period. This outflow is equal to the sediment discharge from the drainage area.
single pier accumulation:	Debris accumulation that occurs only on a single bridge pier as a result of the maximum effective length of the floating debris being less than the effective opening between the bridge piers.
slope (of channel or stream):	Fall per unit length along the channel centerline.
slope protection:	Any measure such as riprap, paving, vegetation, revetment, brush or other material intended to protect a slope from erosion, slipping or caving, or to withstand external hydraulic pressure.
span accumulation:	Debris accumulation that accumulates across an entire span of a bridge structure as a result the length of floating debris exceeding the effective opening between piers.
spill-through abutment:	A bridge abutment having a fill slope on the streamward side.
spur:	A permeable or impermeable linear structure that projects into a channel from the bank to alter flow direction, induce deposition, or reduce flow velocity along the bank.
spur dike:	See guide bank.
stable channel:	A condition that exists when a stream has a bed slope and cross section which allows its channel to transport the water and sediment delivered from the upstream watershed without aggradation, degradation, or bank erosion (a graded stream).
stage:	Water surface elevation of a stream with respect to a reference elevation.
stream:	A body of water that may range in size from a large river to a small rill flowing in a channel. By extension, the term is sometimes applied to a natural channel or drainage course formed by flowing water whether it is occupied by water or not.

streambank erosion: Removal of soil particles or a mass of particles from a bank surface due primarily to water flow in the channel. Other factors such as weathering, ice and debris abrasion, chemical reactions, and land use changes may also directly or indirectly lead to bank erosion.

streambank failure: Sudden collapse of a bank due to an unstable condition such as due to removal of material at the toe of the bank by scour.

streambank protection: Any technique used to prevent erosion or failure of a streambank.

structural countermeasure: A structural component used to prevent, delay or reduce the severity of stream or river problems.

substructure: The components of a bridge which includes all elements supporting the superstructure. Its purpose is to transfer the loads from the superstructure to the foundation soil or rock.

superstructure: The entire portion of a bridge structure which primarily receives and supports traffic loads and in turn transfers these loads to the bridge substructure.

subcritical, supercritical flow: Open channel flow conditions with Froude Number less than and greater than unity, respectively.

thalweg: The line extending down a channel that follows the lowest elevation of the bed.

toe of bank: That portion of a stream cross section where the lower bank terminates and the channel bottom or the opposite lower bank begins.

toe protection: Loose stones laid or dumped at the toe of an embankment, groin, etc., or masonry or concrete wall built at the junction of the bank and the bed in channels or at extremities of hydraulic structures to counteract erosion.

turbulence: Motion of fluids in which local velocities and pressures fluctuate irregularly in a random manner as opposed to laminar flow where all particles of the fluid move in distinct and separate lines.

underflow jam: A type of debris jam that exists near the bankfull level. These types of jams usually occur in small watersheds where the tree height is greater than the channel width.

uniform flow: Flow of constant cross section and velocity through a reach of channel at a given instant. Both the energy slope and the water slope are equal to the bed slope under conditions of uniform flow.

velocity: The rate of motion of a fluid in a stream or of the objects or particles transported therein, usually expressed in m/s (ft/s). The average velocity at a given cross section is determined by dividing discharge by cross-sectional area.

Waterway opening   Width (area) of bridge opening at (below) a specified stage, measured
width (area):   normal to the principal direction of flow.

CHAPTER 1 – INTRODUCTION

1.1 PURPOSE

The purpose of this Federal Highway Administration (FHWA) Hydraulic Engineering Circular (HEC) is to provide information on debris accumulation, guidelines for analyzing impacts associated with debris accumulation, and design guidelines for selecting debris control countermeasures. The design guidelines are based on countermeasures that have been implemented by federal, State, and local transportation agencies at culvert and bridge structures.

1.2 BACKGROUND

Debris accumulation at culvert and bridge structures openings is a widespread problem. The accumulation of debris at inlets of highway culverts and bridge structures is a frequent cause of unsatisfactory performance and malfunction. This accumulation may result in erosion at culvert entrances, overtopping and failure of roadway embankments and damage to adjacent properties, increased local scour at piers and/or abutments, and the formation of pressure flow scour. Consideration of debris accumulations and the need for debris-control structures should be an essential part of the design of all drainage structures.

Structural and non-structural measures have been used effectively to prevent or reduce the size of debris accumulations at bridges and culverts. Structural measures can include features that: (a) intercept debris at or upstream of a structure inlet; (b) deflect debris near the inlet; or (c) orient the debris to facilitate passage of the debris through the structure. Non-structural measures include management of the upstream watershed and maintenance. This document provides measures for both culvert and bridge structures. The measures available for culverts are based on the information included in earlier editions of this manual. Selection of a certain debris countermeasure depends upon the size, quantity, and type of debris, the potential hazard to life and property, the costs involved, and the maintenance proposed.

1.3 DOCUMENT ORGANIZATION

This HEC is organized to provide the following:

- Summarize the various types of debris and the problems associated with debris accumulation at culvert and bridge structures (CHAPTER 2)

- Provide a general procedure for estimating the volume of floating debris upstream of a bridge site, the potential for the debris to accumulate on a bridge structure, and the potential maximum size of the debris accumulation (CHAPTER 3).

- Provide general guidelines for analyzing and modeling debris accumulations on a bridge structure to determine the impacts the debris would have on the water surface profile through the bridge structure and the hydraulic loading on the structure (CHAPTER 4).

- Summarize (describe) the various types of debris countermeasures available for culvert and bridge structures (CHAPTER 5).

- Provide general criteria for selection of debris countermeasures for culvert and bridge structures, and provide design guidelines for structural countermeasures for which guidelines are available (CHAPTER 6).

- Provide general information on maintenance practices (CHAPTER 7).

- Provide references and source materials that provide additional, more comprehensive information on debris and debris issues. The references are grouped alphabetically, by author.

- Provide a synopsis of a survey of State Department of Transportation and American Association of State Highway Transportation Officials (AASHTO) debris issues and mitigation practices. The survey was conducted as part of the effort to update this document.

1.4   DUAL SYSTEM OF UNITS

This edition of HEC-9 uses dual units, SI metric and English. The "English" system of units as used throughout this manual refers to U.S. Customary (CU) units. **An explanation of the metric (SI) unit of measurement is provided in Appendix A. This appendix also contains conversion factors, physical properties of water in the SI and CU units, sediment particle size grade scale, and some common equivalent hydraulic units.**

This edition uses the meter (m) or foot (ft) for the unit of length; kilogram (kg) or slug for the unit of mass; Newton (N) or pound (lb) for the unit of weight/force; Pascal (Pa, N/m^2) or pounds per square foot (lb/ft^2) for the units of pressure; degree Centigrade (°C) or Fahrenheit (°F) for the unit of temperature. The unit of time is the same in both systems. The value of some of the common engineering terms used in this reference in SI units and their equivalent English Units are given in Table 1.1.

Table 1.1. Commonly Used Engineering Terms in SI and CU Units.

Term	SI Units	English Units
Length	1 m	3.28 ft
Volume	1 m^3	35.31 ft^3
Discharge	1m^3/s	35.31 ft^3/s
Acceleration of Gravity	9.81 m/s^2	32.2 ft/s^2
Unit Weight of Water	9810 N/m^3	62.4 lb/ft^3
Density of Water	1000 kg/m^3	1.94 slugs/ft^3
Density of Quartz	2647 kg/m^3	5.14 slugs/ft^3
Specific Gravity of Quartz	2.65	2.65
Specific Gravity of Water	1	1
Temperature	$^\circ$C = 5/9 ($^\circ$F − 32)	$^\circ$F

(page intentionally left blank)

CHAPTER 2 – DEBRIS CHARACTERIZATION

2.1 DEBRIS CLASSIFICATION

Flood flow reaching a culvert or bridge structure typically carries floating as well as submerged debris. As discussed in further detail in the next section of this manual, debris should be a concern to highway engineers because it can accumulate at and obstruct the waterway entrance of culverts or bridges, adversely affect the operation of the structure or cause failure of the structure.[14,17,40,47,49,50] A thorough study of the supply of debris originating in the drainage basin is essential for proper design of a drainage structure.

2.1.1 Types of Debris

The selection of a debris countermeasure depends on the type of debris transported to the site; therefore, the various types of debris should be defined and classified to assist in the selection of an effective debris countermeasure. This current edition retains, but slightly modifies, the classification system used in earlier editions. This classification is presented as follows:

Very Small Buoyant Debris or No Debris.

Small Floating Debris. Small floating debris includes small limbs or sticks, orchard prunings, tules, leaves, and refuse. This material can be easily transported by the stream and overland flow. Therefore, this type of debris can be introduced into the stream from the local runoff from a watershed, and then easily transported downstream by the stream flows. This type of debris also comes from tree and vegetation that are introduced into the stream due to bank erosion, landmass failures, wind action or collapsing due to biological factors such as decay and old age, or from the loss of foliage during the changing of seasons. There are usually no significant problems associated with this type of debris at bridge structures; however, it is an important component in the development of mature debris jams of large floating debris, and it could accumulate at and clog culvert structures.

Medium Floating Debris. Medium floating debris consists of tree limbs or large sticks. The source of this material comes from trees introduced into the stream by bank erosion, mass wasting, windthrow, or collapsing of trees due to ice loading, beaver activities, or biological factors such as old age and diseases; or from erosion of emergent and riparian trees within the streams. Vegetation within the floodplain could also be a source of this type of debris. This type of debris could accumulate at both culvert and bridge structures.

Large Floating Debris. Large floating debris consists of logs or trees (such large floating debris is also known as "drift"). The sources of this type of debris are the same sources discussed for "Medium Floating Debris". Transport and storage of this material depends on discharge, channel characteristics, the size of the drift pieces relative to the channel dimensions, and the hydraulic characteristics (depth and slope) of the system.[17] In small and intermediate size streams, this material is not easily transported, and it is usually transported during larger floods or prolonged periods of high water.[17] Once introduced into the main channel of these small and intermediate size streams, this material can form into a jam, which is a collection of debris formed around large, whole trees that may be anchored to the bed or banks at one or both ends.[17,61] Larger streams and rivers do not store much of this material within the channel. During most flood events, a larger stream will transport nearly all large floating debris entering the reach.[17] The size of the jam depends on the type of vegetation existing within the watershed and the channel characteristics transporting the material. This type of debris causes a significant problem at bridge structures because of its size, shape, and facility for entrapment on bridge piers.

Fine Detritus. Fine detritus consists of silt, sand, and fine gravel more or less devoid of floating debris. The size of this material ranges from 0.004 to 8 mm (0.00016 to 0.31 inches). This type of debris is transported along the bed and in the water column above the bed, i.e., as bed and suspended load. The source of this material is from sheet and rill erosion, gully erosion, landmass movement, and channel and bank erosion. Sediment yield rates for this material can be significantly influenced by the conditions of and changes within the watershed (e.g., urbanization, fire, etc.). Deposition of fine sediment could possibly block a culvert structure or significantly reduce the waterway opening through a bridge structure.

Coarse Detritus. Coarse detritus consists of coarse gravel or rock ranging in size from 16 to 256 mm (0.6 to 10 inches). The source of this material is from bed and/or bank erosion, gully erosion, or landmass movement. This material is usually transported as bed load, however it can be transported as both bed and suspended load within high gradient streams or gullies. Course detritus deposition can easily block a culvert entrance or significantly reduce a bridge waterway opening.

Boulders. Material comprised of large rock ranging in size from 256 to 2048 mm (0.84 to 6.7 feet). This type of material is usually associated with steep mountain streams or gullies, and it is transported as bed load. The source of the boulders is from bed and/or bank erosion or landmass movements. This material can easily block the entrance to a culvert and/or cause damage to the bridge piers from the impact forces.

Flowing Debris. Flowing debris is a heterogeneous fluid mass of clay, silt, sand, gravel, rock, refuse, trees, and/or branches. In general, it is a combination of the different types of debris mentioned above.

Ice Debris. Ice debris is accumulation or transport of ice floes in the waterway. This edition of the document does not describe or characterize this type of debris in any detail. In future editions, FHWA intends to add supplementary information based on results of on-going research efforts.

2.2   FLOW BEHAVIOR OF FLOATING DEBRIS

A brief discussion on the flow behavior of large floating debris is provided in the previous section. A more detailed discussion of the subject is provided in this section because it is an important concept to consider for evaluating the potential for debris accumulations at bridges and/or for developing watershed management plans for debris reduction. The potential for transport, mode of transport, formation of debris rafts, potential locations for debris accumulations, and general characteristics of debris accumulations at bridge structures are discussed in the remaining portions of this section. The information for these various topics was obtained from several different sources[14,17,18,43,59,63], but most of the information was obtained from the report prepared by Diehl.[17]

2.2.1   <u>Debris Transport and Transport Mechanisms</u>

The potential for transport depends on the discharge, channel characteristics, debris source availability, size of the floating debris relative to the channel dimensions, orientation of the debris relative to the channel alignment, and type of anchorage.[17,18,43] The potential for transport increases with increasing discharge due to the increase in the flow velocity, depth, and energy slope of the river.[17] Unfortunately at the present time, in most cases, there are no relationships available that define the minimum velocity and/or slope necessary to initiate transport of large floating debris. However, with respect to ice debris, the AASHTO LRFD Bridge Design Specifications describe several relationships regarding ice pressures and loads.[5] Ongoing research is investigating this and other ice debris issues.

The width, depth, and slope of the channel are important channel characteristics that influence the potential for transport.[17]   In general, this potential increases with increasing widths. Consequently the abundance of floating debris stored in the channel typically decreases with increasing widths. The length of the large floating debris transported increases with increasing channel widths.[17,63]   Narrow streams rarely transport large floating debris, except for steep streams subject to debris torrents.[14,17] For narrow streams, trees and logs, (i.e., large floating debris) are usually longer than the width of the channel, so they typically become lodged across the channel, and rarely move without being broken into smaller pieces.[17]   For most intermediate size streams, only some of the large floating debris is transported during large floods since most of it accumulates within the channel to form sizable debris jams.[17] Furthermore, for rivers and wide streams with adequate flow depth, nearly all of the large floating debris introduced into the main channel is transported by frequent flood events.[14,17] The depth of flow within the channel has to be deep enough to buoy up large floating debris. The depth sufficient to float logs and large trees is about the diameter of the tree butt plus the distance the roots extend below the butt.[17] The potential for transport is higher for high-gradient streams than it would be for low-gradient streams with the same channel dimensions, since the forces of flowing water on stored debris in low-gradient streams are less.[17] Stored debris can be abundant in large, low-gradient channels.[17]

The ratio of the effective length of the tree to the width of the channel is an important factor in a waterway's capacity to transport a particular size of debris.[17] The relation of the length of the debris pieces to the channel width is a primary indicator in defining the transport rate and the type and amount of debris stored in the channel. The potential for debris transport increases as the ratio of the debris length over the channel width decreases. For example, more debris would be transported in an intermediate size stream if the debris length is one-half of the channel width, than would be transported if the debris length equals the width of the channel. The maximum size of the debris is limited by the channel width; however, the amount of debris is limited by the supply of debris and the capability of the channel to transport the debris.

The potential for transport is influenced by the orientation of the debris relative to the channel alignment and type of anchorage.[18] Isolated pieces are more likely to be transported than pieces within debris jams or clumps. Also, trees with the root mass oriented upstream are more likely to be transported than trees with the root mass downstream or near the stream bank since it is easier for the piece of debris to be rotated by the flow.[18] Debris that is anchored to the side and perpendicular to the flow would most likely remain in place and not be transported downstream until the debris is dislodged from the bank due to bank erosion.

After mobilization has occurred, large floating debris is transported either by floating along the water surface or dragging along the bed. Observations noted by Diehl indicate that debris is typically transported on the surface as individual pieces aligned with the flow and traveling at about the same velocity as the average water velocity at the surface.[17] The results of a physical model study performed by Ng[43] agrees with Diehl's observation. Floating debris can occasionally be transported in short-lived clumps that eventually get broken apart due to turbulence.[14,17] The debris typically concentrates in a path occupying only a small fraction of the channel width. This path is defined by the zone of convergence that exists in some channels near the thalweg of the streams where the flow is the deepest and fastest under some flow conditions.[17,43] The zone of surface convergence for a straight and curved channel is illustrated in Figure 2.1.[17] The flow patterns reflected in this figure are hypothetical flow patterns at a particular location in a single bend during bankfull flow conditions, and the flow patterns would be entirely different for larger flood flows, different radii of curvature, or different channel conditions.

Observations noted by Chang[14] indicate that floating debris within straight reaches tends to move inward to the thalweg at the rising stage of a flood and outward to the banks at the receding stage. The reasons are unclear; however, it could possibly be related to changes in the direction of the secondary flow patterns within the channel. As noted by Ng[43], the opposite pattern can occur, i.e., outward to the banks during the rising and inward to the thalweg during the receding, if water leaves the channel and flows into the floodplains. Diehl's[17] observations in curved channel reaches indicate that floating debris may be transported on the outside of the curve during both rising and falling conditions. He also observed that debris typically travels between the center of the channel and the outside bank rather than in contact with the bank vegetation.

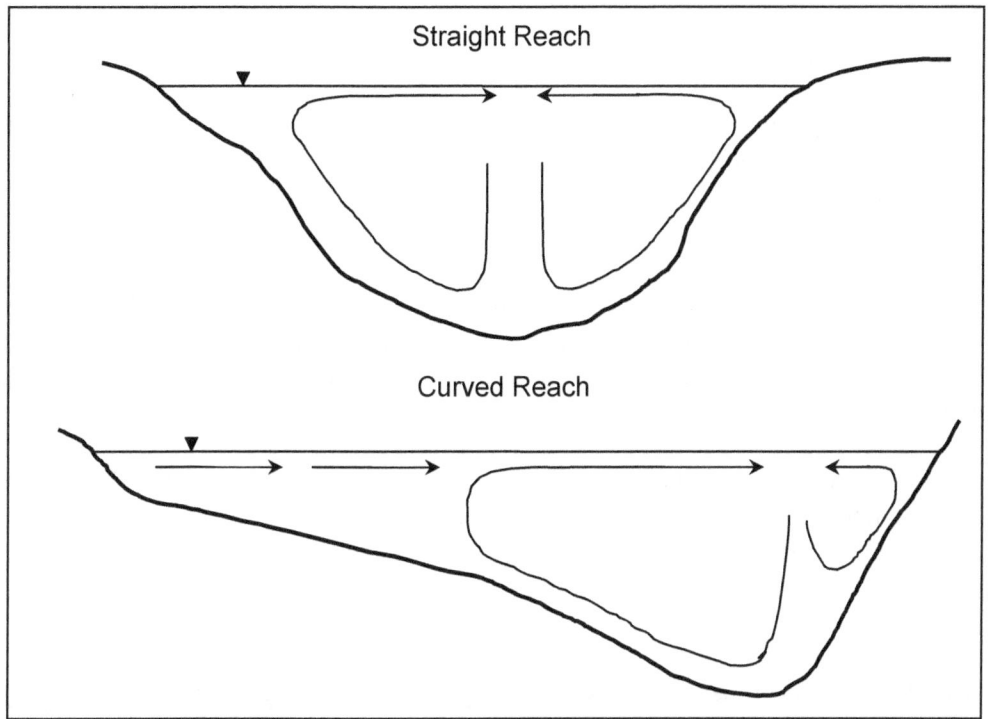

Figure 2.1. Hypothetical Patterns of secondary flow in straight and curved channels.

2.2.2 Debris Jams and Debris Dams

Floating debris introduced into the channel can form into debris jams or dams.[17,63] Debris jams are usually formed when large, whole trees are introduced into the channel and are anchored to the bed or banks at one or both ends. The large trees act as a filter by trapping smaller floating debris and possibly sediment. The size and location of a jam depend on the size of the stream and the size of the trees. In small streams, a fallen tree may not be readily transported. However, this tree may trap and accumulate smaller debris from upstream and form a debris jam. Conversely, larger rivers can readily transport the debris downstream so it may not be able to accumulate into a large jam. Most of the accumulations in large rivers occur on the channel margins or outside the channel on islands, in floodplain forests, and in sloughs.[17,63]

As noted by Wallerstein and Thorne[63], debris jams influence the geomorphology of rivers by influencing the overall channel form (i.e., they distort pool-riffle sequence and gravel bar formation); by changing the channel topography (i.e., they influence the erosional and depositional processes and widen the channel through bank erosion); and by increasing the channel apparent roughness through increased energy dissipation and eddy formation. (Note energy dissipation cannot be increased in a river, it can only be redistributed. Instead of the energy being dissipated along the channel, it is dissipated in intense drops over and around the debris. Upstream, the damming effect reduces the dissipation along the backwater-affected reach).

Wallerstein and Thorne[63] classified debris jams according to what they called "engineering-geomorphic impacts" as follows:

Underflow Jams occur in small watersheds where the fallen trees span the channel at bankfull level (define bankfull level in glossary). The in-channel geomorphic impact associated with this type of debris jams is minimal; however, local bed scour could occur under the jam during high flows.

Dam Jams usually occur when the tree height is approximately equal to the channel width. These types of jams can cause significant localized bank erosion and bed scour due to the constriction in flow, and backwater effects upstream that could cause sediment deposition upstream of the jam.

Deflector Jams usually occur when the channel width is slightly greater than the average tree height. They usually redirect the flow to one or both of the banks causing bed and bank erosion that could result in more trees being introduced into the river. They can also create backwater sediment wedges and downstream bars depending on the level of dissipation caused by the jam.

Parallel Jams exist when the channel width is significantly greater than the debris length, and the flow is capable of rotating the debris so that it is parallel with the flow. Bank erosion and bed scour associated with these jams are usually minimal. Parallel jams could actually stabilize the bank toe and protect it from erosion, and they may also initiate or accelerate the formation of mid-channel and lateral bars.

2.2.3 Debris Accumulation

Floating debris can accumulate at various locations and at obstructions within the river such as bridge piers and abutments, mid-channel bars, point bars, island heads, the streambed, or in pools along the base on the outside bank of bends. Debris accumulations typically grow in the upstream direction through the accretion of additional floating debris and fine and coarse sediment.[17] The rate of accumulation depends on the concentration, defined as number of debris per length of channel, of floating debris that is being transported and the magnitude of the flood.[14] In general, debris accumulations occur most frequently and in the largest sizes where the path of floating debris encounters obstructions.[17] The potential for trapping of debris at a bridge structure can be aggravated by the location and type of bridge piers.[14,17] Multiple columns can act as a sieve unless exactly aligned with flow; however, alignment of piers to flow at all flood levels for which debris is transported is unlikely. The gaps between columns are narrow relative to length of the floating debris, resulting in a high potential for accumulating debris. Floating debris can become entangled in a group of columns in ways that are not possible for a single-column pier. Floating debris accumulations at bridges generally fall into two classes: single-pier accumulations and span-blockage accumulations.[17]

Single-pier accumulation occurs when the maximum effective length of the floating debris is less than the effective opening between the bridge piers. The effective length of debris is the length of the debris element that can support the load of the debris accumulation. The effective opening corresponds to the distance between the piers normal to the approaching flow. The width of the opening can be determined by extending lines parallel to the approaching flow upstream from the nose of each pier and measuring the perpendicular distance between the two lines. As noted by Diehl, single-pier accumulations typically contain one or more trees extending the full width of the accumulation perpendicular to the approaching flow.[17] The full-width tree can be either at the surface or submerged and concealed beneath smaller floating debris. Pier placement is extremely important for this type of accumulation. Even if the span length is significantly greater than the maximum length of the floating debris, a pier located within the path of floating debris (Figure 2.2) can result in a high potential for accumulation at the pier.[17]

Figure 2.2. Single pier debris accumulation (led to pier scour failure).

Span-blockage accumulations occur when the length of floating debris exceeds the effective opening between piers, resulting in the floating debris resting against two piers (Figure 2.3). This type of accumulation can also exist between a pier and an abutment. A similar type of accumulation can occur between a pier and a bank or other large fixed object, such as boulders and trees that can support one end of the floating debris. Like the potential for single-pier accumulations, the potential for span blockages is influenced by pier placement.[17]

Figure 2.3. Span blockage accumulation bridge failure. (Louisiana)

As noted by Diehl, most large debris accumulations are similar in shape. Floating debris is initially trapped on the pier perpendicular to the approaching flow, but as the accumulation increases in size, debris accumulates parallel to the upstream edge of the accumulation. This process results in an accumulation with a curved upstream edge, and with the upstream nose of the accumulation raft near the thalweg, where most of the debris is transported. The accumulation is typically deepest at the piers that support them, and widest at the surface. The potential to achieve a roughly rectangular cross section from the bed to the water surface depends on the abundance of debris, prolonged periods of high water, or multiple floods without removal of the accumulation.[17]

Debris accumulations initially form at the water surface, grow toward and eventually become part of the streambed. As the water surface increases during a flood, floating debris already existing on the bridge usually remains in place as additional floating debris accumulates at the water surface. When the flood subsides, the new accumulated debris usually slides downward until it rests on the bed or on the previous debris accumulation to form a solid mass with irregular protrusions around the base of the pier.

2.3   PROBLEMS ASSOCIATED WITH DEBRIS

There are various potential problems associated with debris accumulations. In general, debris accumulations can adversely impact the conveyance through a culvert or bridge structure, exacerbate the contraction and local scour at a bridge structure, increase the hydraulic loading on a bridge structure, and cause upstream flooding. Several failures of highway bridges, roadway embankments and highway culverts have been attributed at least in part to debris accumulation. Figure 2.4 and Figure 2.5 shows two bridge failures attributed to debris accumulation during flood events.[49, 53]

Figure 2.4. Missouri Highway 113 bridge over Florida Creek near Skidmore, Missouri.

Figure 2.5. Debris accumulation failure at bridge located in Oklahoma.

Debris accumulation can partially or totally block the waterway opening for a culvert or bridge structure.[14,17,40,47,49,50] A massive accumulation of woody debris at a bridge is shown in Figure 2.6. Blockage of large portions of the waterway opening will increase backwater elevations upstream, increase flow velocity through the contracted opening under the structure, and modify flow patterns.[14,17,49,50] The increase in backwater upstream significantly increases the upstream inundation boundaries. High velocity contracted flow and large water surface elevation differences from the upstream to downstream side of the bridge can cause high drag and hydrostatic forces on the structure that can cause structural failure and collapse. Flows increased in velocity by the obstruction of the waterway and deflected away from the main channel can cause severe erosion near abutments or stream banks. Reduction in bridge waterway opening by debris can also cause a reduction in the flow rate required to overtop and potentially damage bridge approach roadways and embankments. Large accumulations could adversely affect the flow patterns near the structure by creating a strong lateral flow across the river towards the adjacent piers or embankment fill at unanticipated and potentially severe angles of attack, resulting in deep local scour at piers or abutment embankment fill. [17,50,51]

Figure 2.6. Debris accumulation at a bridge structure.[17]

Debris accumulations can also exacerbate the scour near the culvert or bridge structure. Figure 2.7 depicts streambank failure associated with a span blockage (already shown in Figure 2.3). As stated above, the blockage of flow area from debris accumulations can cause a significant increase in the flow velocities through the bridge structure. This increase in flow velocities and boundary shear stresses may cause an increase in the contraction scour through the bridge if the entire bridge opening is affected.

Figure 2.7. Debris induced bridge and bank failure structure. (Louisiana)

A laboratory study performed by Dongol[20] showed that debris accumulations (simulated using cylindrical shaped, PVC disks) cause larger and deeper scour holes to develop as a result of the significant increase in the downward velocity below the debris and the increase in both the horseshoe vortex size and the contact area of the vortex. The increase in both the contraction and local scour near the bridge structure could possibly damage or cause failure of the structure due to undermining of the pier footing or the abutment toe. Unfortunately, there has been only limited research conducted on local scour at piers with debris accumulation. Therefore, the scour associated with debris accumulation is extremely difficult to assess with any reliability.

Damage and failure of several bridges has been related to the increase in the hydraulic loading on the structures caused by debris accumulations.[14,17,40,47,49,50] Highway bridges partially or fully submerged during a flood event are subjected to various types of forces. These forces include hydrodynamic drag and side forces, hydrostatic forces, buoyant forces, hydrodynamic-lift forces, and impact forces.[49]

Hydrodynamic drag forces result from the reaction of the water as it flows around an object, and it acts parallel to the direction of flow.

Side forces are similar to drag forces, but act perpendicular to the flow direction.

Hydrostatic forces on the bridge elements are related to the differential in water surface elevations at the upstream and downstream sides of the structure caused by the flow constriction through the bridge.

Buoyant forces result from the displacement of water by the bridge or by the debris lodged under the bridge.

Hydrodynamic-lift forces are related to the total dynamic pressure force acting in the vertical direction perpendicular to the flow direction and the side force.

Impact forces are related to the moving debris colliding with the bridge structure.

Debris accumulations cause an increase in these forces due to the increase in upstream water surface elevations, increase in the flow velocities through the bridge, and increase in projected area of these forces on the structure.[49] Increases of these forces may cause the bridge structure to collapse either by buckling of the bridge substructures, shearing of roadway deck supports, or overturning of the structure (see Figure 2.8).[49,50]

Figure 2.8. Effects of debris accumulation at a bridge structure. (New York)

Miscellaneous problems associated with debris accumulation include difficult and expensive maintenance programs required for debris removal, an increase in fire potential near the structure, and minor damage to the structure.[14]

(page intentionally left blank)

CHAPTER 3 – ESTIMATING DEBRIS QUANTITIES

3.1   INTRODUCTION

The information presented in this Chapter focuses mainly on floating debris. This chapter does not address sediment yields and transport rates for fine and coarse sediment that is thoroughly documented in several references.[24,34,38,54,58,65] Most of the information presented in this chapter is based on a detailed study conducted by the U.S. Geological Survey (USGS), in cooperation with the FHWA. The study included an analysis of data from 2,577 reported debris accumulations and field investigations of 144 debris accumulations. Guidelines for the assessment of debris potential in the form of a detailed assessment method was proposed as a result of the study by Diehl.[17] The use of these guidelines requires familiarity with the specific regional characteristics of the local stream morphology and debris loading characteristics. As in all aspects of river problems, familiarity with the historical land-use activities, geologic and climactic conditions and the way in which these factors affect stream morphology and the debris loads in streams is imperative for making effective management decisions about debris production.

The evaluation of debris accumulation on bridges has been separated into three major phases by Diehl:[17]

1) Estimate the potential for debris delivery to the site,
2) Estimate the debris accumulation potential on an individual bridge element, and
3) Calculate hypothetical accumulations for the entire bridge.

These phases can be further subdivided into eight tasks (Table 3.1). Each of these three phases is discussed in detail, even though the last phase is based on a qualitative methodology. It is presented because it is the most thorough information available on the subject matter. **However, caution and a familiarity with the specific regional characteristics of the local stream morphology and debris loading characteristics should be used when applying this information.** Simple examination of debris accumulations by state DOT's during maintenance for important, but easily measured parameters outlined in the following procedure would provide specific local information necessary for improved future debris accumulation estimates. Although debris problems are widespread, conditions and parameters controlling debris production are specific to the local watershed conditions.

As indicated in Table 3.1, estimating the volume of large debris within the watershed is not a major phase required in evaluating the potential for debris accumulations at a bridge. This information, however, is beneficial in developing a better understanding of the debris dynamics within the watershed, provide an indication of the overall debris conditions within the watershed, and provide an indication on the potential for debris being delivered to a structure within the watershed. Therefore, Diehl's and Bryan's[18] procedure for estimating the volume of large debris within the watershed is also presented within this chapter.

Table 3.1. Major Phases and Tasks in Evaluating Debris Accumulation Potential at a Bridge.

Major Phase	Task
1. Estimate potential for debris delivery	a. Estimate potential for debris delivery to the site. b. Estimate size of largest debris delivered. c. Assign location categories to all parts of the highway crossing.
2. Estimate debris potential on individual bridge elements (i.e., piers, abutments, etc.)	a. Assign bridge characteristics to all immersed parts of the bridge b. Determine accumulation potential for each part of the bridge
3. Calculate hypothetical debris accumulations for the entire bridge	a. Calculate hypothetical accumulation of medium potential b. Calculate hypothetical accumulation of high potential c. Calculate hypothetical chronic accumulation
Source: (17)	

3.2 DEBRIS VOLUMES

The volume of large debris within a watershed can be determined using a procedure applied by Diehl and Bryan[18] for a basin of the West Harpeth River in Tennessee. The general procedure involves selecting several different reaches of the river that are representative of the conditions upstream and downstream of the selected reach. The representative reaches could be selected using aerial photographs and/or during the reconnaissance field investigation. Debris greater than a certain length is counted and measured within each of the reaches. Debris concentration is then calculated for each of the reaches as either:

- the cubic meters of debris per kilometer of channel (e.g., 27 m^3/km for a reach that is 3 kilometers in length and contains 81 m^3 of debris), or

- the number of pieces within a height range per kilometer of channel (e.g., 25 (10 meters long)/km for a reach that is 3 kilometers in length and contains 75 pieces of 10 meter long debris).

The total volume of debris for each of the individual reaches is then estimated by multiplying the debris concentration by the total length that the selected reach represents. For example, a reach that has a debris concentration of 27 m^3/km (i.e., 3 kilometers in length containing 81 m^3 of debris) and a total representative length of 12 km would have a total volume of debris of 324 m^3 (i.e., 27 m^3/km times 12 km equals 324 m^3).

Finally, the volumes for each of the individual reaches are summed to determine the total volume of debris. For example, a watershed that has three reaches with individual debris volumes of 324 m^3, 500 m^3, and 210 m^3 would produce a total volume of debris of 1,034 m^3 (i.e., 324 m^3 plus 500 m^3 plus 210 m^3 equals 1,034 m^3).

During the counting and measuring of the debris, additional information about the debris should be noted and documented. The information should include:

- The length of the piece, measured from root mass or spar butt to spar top;

- The diameter at the butt and at the top of the piece;

- If it has a straight or curved stem;

- The abundance of branches, as in none, few, or many;

- If the bark is present or not;

- The condition of root mass, as in dirty, one-sided, symmetrical, worn, and/or gone;

- The orientation within the channel, such as parallel, perpendicular, or diagonal to channel alignment;

- The position in channel, such as on the bed, on the bank, or on the bed and bank; and

- The type and extent of anchorage, such as on top of bed, entrenched in bed, tangled in vegetation, or part of a debris pile.

3.3   POTENTIAL FOR DEBRIS DELIVERY

The first phase in evaluating the potential for debris accumulation at a bridge is to estimate the potential for debris delivery to the bridge site. The tasks involved for this phase include estimating the potential for delivery of floating debris, estimating the largest size of floating debris delivered to the site, and assigning location categories to all parts of the highway crossing.

3.3.1   Task A: Estimating Potential for Debris Delivery to Site

The potential for debris delivery is evaluated based on the potential for the debris to be transported downstream to the bridge site and the potential for debris generation as defined by direct and indirect evidence. Observations of floating debris provide the most direct evidence for assessing the potential for debris delivery. These observations could be made of the channel system or of accumulations at bridges and/or at other sites upstream of the bridge structure or within a basin of similar characteristics. Even though present observations indicate that there is a low potential for debris delivery, infrequent catastrophic events or changes in the watershed could still result in abundant floating debris in the future. Therefore for such events, indirect evidence should be considered.

Direct evidence for high delivery potential includes the following observations:

- Multiple cases of floating debris accumulation at bridges.

- Chronic floating debris accumulation at one or more sites.

- Floating debris accumulation at sites where potential for accumulation would be low if floating debris were not abundant.

- Abundant floating debris stored in the channel.

- Past need for debris removal in the channel system or at bridges.

Direct evidence of low potential for drift delivery may be indicated by the following observations:

- Negligible floating debris delivered in major events, especially at sites with a high potential for trapping floating debris or at typical debris-accumulation sites.

- All of the floating debris accumulates in forested channel upstream.

- Floating debris in the channel is stationary during floods because of low flow velocity.

The potential for debris delivery can also be assessed from indirect evidence of debris generation. As previously discussed, a major source of floating debris is from bank erosion. Therefore, evidence of existing or potential bank erosion can be considered as indirect evidence for high potential of debris generation. Observations of indirect evidence for abundant debris generation include:

- Widespread bank erosion in the upstream channel system.

- History of changes in the upstream channel system, including degradation, lateral migration, widening, channelization, in-stream gravel mining, widespread drainage, or dams.

- Prospects of changes in the channel system.

- Hydraulic and geomorphic factors indicating stream instability.

- Widespread timber harvesting in the basin.

- History or prospect of marked changes in basin land use.

- In-stream gravel mining.

Indirect evidence for low potential of debris generation includes the following observations:

- The inability of woody vegetation to grow along the channel system and on steep slopes leading down to the stream channels.

- The channel system is stable and is unlikely to experience any significant change.

Where indirect evidence indicates that there is a high potential for existing or potential future debris generation, the ability of the channel system to transport the debris will control the potential of the debris delivery to the site. In general, most streams are capable of transporting some of the debris, and one should assume that the stream is capable of transporting the debris unless there is evidence to the contrary. Stable, densely forested streams transport little debris and can be assumed to have a low delivery potential as long as the forest will not be cleared in the future.[17]

Applying the information above, a "High Delivery Potential" exists when there is an abundant amount of direct evidence of debris delivered to the site, or there is indirect evidence of existing or future debris generation within the watershed and the upstream channel is capable of transporting the floating debris to the site; and, a "Low Delivery Potential" exists when there is a sparse amount of direct evidence of debris delivered to the site and there is no existing indirect evidence of future debris generation within the watershed, or when the upstream channel is incapable of transporting the floating debris to the site.

3.3.2   Task B: Estimating the Largest Debris Size Delivered to Site

The second task of the first phase is to estimate the size of the largest debris delivered to the site (Maximum Design Log Length). This debris size influences the potential size of the debris accumulation. The largest debris delivered to the site is influenced by the channel dimensions upstream of the site, particularly the channel width. These dimensions may change over the project life of a bridge as a result of future stream instabilities, and these changes should be accounted for when defining the channel dimensions.

As illustrated in Figure 3.1, the maximum design log length is estimated on the basis of the narrowest channel width immediately upstream from the site. This distance should be measured perpendicular to the banks or lines of permanent vegetation at the inflection points between bends.

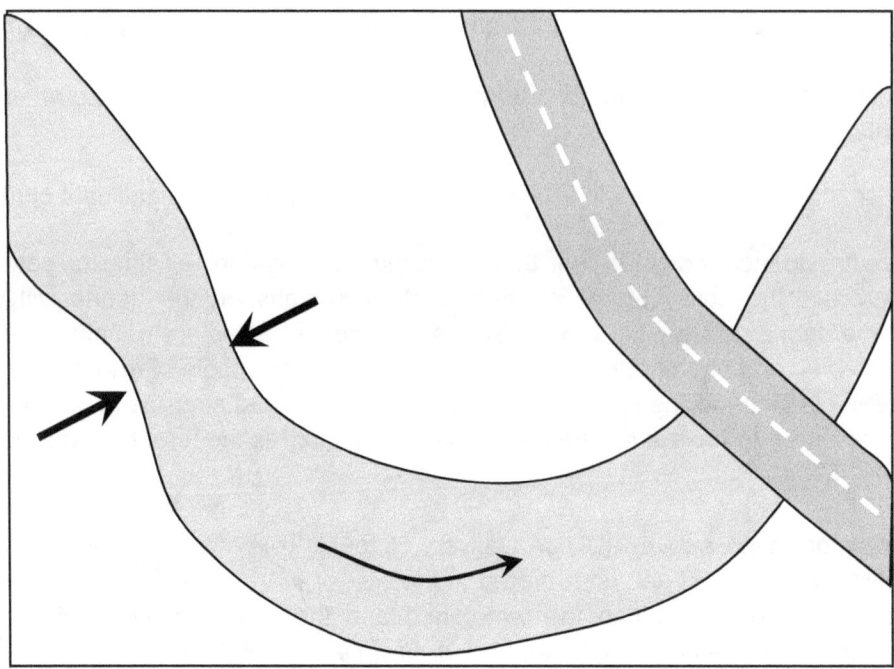

Figure 3.1. Schematic Illustrating Estimate of Maximum Design Log Length

The minimum channel depth required to transport large trees is estimated to be about the diameter of the butt plus the distance the root mass extends below the butt, or roughly 3 to 5 percent of the estimated tree length.

The design log length represents a length above which logs are insufficiently abundant and insufficiently strong throughout their full length to produce an accumulation equal to their length, and it does not represent the absolute maximum length of trees within the watershed upstream of the site. Diehl[17] recommends estimating the design log length at a given site as the smallest of these three values:

- Width of the channel upstream from the site.

- Maximum length of sturdy logs. The height and diameter of mature trees on the banks determine the maximum length of the logs that can be delivered to the bridge as floating debris and capable of withstanding hydraulic forces when lodged against the piers. The maximum sturdy-log length seems to reach about 24 m (about 80 ft) in much of the United States; however, it may be as long as about 45 m (about 150 ft) in parts of northern California and the Pacific Northwest

- 9 m (30 ft) plus one quarter of the width of the channel upstream from the site, in much of the United States. As indicated by Diehl[17], this third constraint reflects the rarity of long logs and their breakage during transport, and it should not be considered for sites located in northern California or the Pacific Northwest.

3.3.3   Task C: Assigning Location Categories to All Parts of the Highway Crossing

The last task for the first phase involves assigning location categories to all parts of the highway crossing. As previously mentioned, debris is generally transported along a relatively small portion of the channel cross section. As a result, some areas of a site may be entirely free of debris transport, whereas other areas may receive a significant amount of debris. The various categories include:

Sheltered Location. A sheltered location is defined for the section of the bridge that includes a forest area directly upstream of the bridge that traps the transported debris and prevents it from being delivered to the bridge. This category should only be applied when the gaps between trees are much narrower than the average tree height and the width of forest along the direction of flow is more than a double line of trees. Intuitively, this category should not be applied to the upstream forest area if it is potentially subject to clearing.

Bank/Floodplain Location. This category includes the slope of the bank, top of the bank, and the floodplain since piers located on the slope of banks or at the top of the bank are just as likely to accumulate debris as piers located in the floodplain. The floodplain includes any area outside of the channel that is inundated in the design flood to a depth sufficient to transport drift, and it may be either clear of trees or a forested area that is subject to future clearing. If there is evidence that debris is transported within the slope of the channel banks, then the banks should not be assigned to this category.

In the Channel Location. Debris can be transported anywhere in the channel. As expected, debris accumulations are more common for "in the channel" locations than for "bank/floodplain" locations, so the potential for debris accumulations for this category is higher than for the previous category. In humid regions, the "in the channel" location is typically defined by the base flow. In arid regions, where the base flow is relatively low, or for ephemeral streams, this location is typically defined between the toes of the banks. If there is evidence that debris is transported within the slope of the channel banks, then they should be assigned to this category.

In the Path Location. This category is defined for the portion of the cross section in which the majority of the debris is transported. As previously mentioned, floating debris is generally transported in most streams along a relatively narrow path within the channel where the secondary circulation currents converge at the surface. In a straight reach, this convergence zone typically coincides with the thalweg of the channel where the flow is the deepest and fastest. In a curved reach, this zone generally exists between the thalweg and the outside bank of the bend. The best way to identify the debris path is to observe it during bank-full or high flow conditions. The observations do not need to be of large pieces of debris since all floating material responds similarly to the flow pattern. If observations indicate that the debris path includes part of the bank or part of the flood plain, then they should be assigned to this category. If high-flow observations are not available, observations during base flow can confirm the estimates based on channel characteristics. If direct observation is impossible for all flow conditions, then the location of the debris path can be estimated based on channel characteristics and assuming that the width of the debris path is about one-third the channel width. If the location of the debris path is indefinite, then several different locations of the debris path could be considered, i.e., the left third of the channel, the middle third, or the right third, or the entire channel could be assigned to this category, which would reflect the worst case scenario.

3.4   POTENTIAL FOR DEBRIS ACCUMULATION

The second phase in evaluating the potential for debris accumulation at a bridge is to estimate the debris potential on individual bridge elements. The tasks involved for this phase include assigning bridge characteristics to all immersed parts of the bridge and determining the accumulation potential for each of these parts.

3.4.1   <u>Task A: Assigning Bridge Characteristics to All Immersed Parts of the Bridge</u>

There are certain characteristics of a bridge structure that influence the potential for debris accumulation. So, the bridge structure should be divided into different components and the potential for accumulation should be evaluated separately for each of the components. The different components include piers, abutments, any gaps between fixed elements of the bridge opening, and the portion of the superstructure submerged during the design flood event.

An effective width needs to be determined for both horizontal and vertical gaps in the bridge structure below the design water surface elevation. Horizontal gaps between adjacent piers, between each bank and the nearest pier in the channel, and between each abutment and the nearest pier in the channel are common locations for large accumulations. The potential for accumulation is high when the effective width of the horizontal opening is less than the length of the longest piece of debris delivered to the bridge. When this is the case, debris typically comes into contact with one of the bridge elements, and then rotates downstream until it becomes lodged against another of the bridge elements. The effective width of the horizontal gaps should be reduced to account for any skew in the bridge to the approaching flow.

When the water surface elevation is at or above the bottom elevation of the superstructure ("low chord elevation"), debris can become trapped vertically between the superstructure and the streambed below it. When floating debris hits the superstructure, most of the pieces rotate to one side and remain at the water surface, resulting in an accumulation against the superstructure at the surface. However, some debris could be lodged between the streambed and the superstructure as a result of the upstream end of the debris rotating downward until it encounters the streambed after striking the superstructure roughly endwise. The height of the vertical gap between the low chord elevation of the bridge and the elevation of the streambed beneath should be based on the minimum height since the height most likely will vary along the bridge due to the changes in the elevations of the low chord and/or the streambed.

Narrow openings of the structure elements of the bridge at the water surface elevation also determine whether debris would be deflected or trapped. Piers and superstructures with narrow openings that convey flow are significantly more likely to trap and accumulate debris. Examples of such structures include:

- Multiple closely spaced pier or pile groups,

- Closely spaced rows of piers,

- Exposed pier footing piles,

- A pile bent or a pier made of a single row of columns skewed to the approaching flow,

- Open truss superstructures,

- Superstructures with open parapets of pillars and rails, and

- Various types of connections between the pier caps and the bridge deck.

Also, any pier with existing accumulations should fall under this classification. Elimination of these narrow openings by using a single solid pier (wall, cylinder, or hammerhead), a superstructure with a solid parapet, and a solid beam that is connected directly to the pier would increase the likelihood of the debris being deflected and not trapped by the structure.

3.4.2   Task B: Determining Accumulation Potential for Each Bridge Component

The first step in determining the accumulation potential for each of the bridge components is to assign a location category described in the previous section to each of the components. The selected category for a horizontal gap should be based on the most debris-prone location category occupied by the fixed elements that define the gap. For example, a horizontal gap from a bank or abutment to a pier located in the debris path should be assigned to the "in the path" location category, or a gap that has one of the fixed elements sheltered while the other element in the floodplain should be assigned to the "floodplain" location category.

After the categories have been assigned to each of the components, the potential for debris to span a horizontal or vertical opening between fixed elements of the bridge can be estimated using the information provided in Table 3.2, and the potential for accumulation on each of the piers can be estimated using the information provided in Table 3.3.

3.9

Table 3.2. Determining Potential for Debris Accumulation across a Span or Vertical Gap.

Accumulation Potential	Gap Wider than Design Log Length	Location Category	Potential for Debris Delivery
Low	-	Sheltered	-
	Yes	-	-
	No	Bank/Floodplain	Low
	No	In the Channel	Low
Medium	No	In the Path	Low
	No	Bank/Floodplain	High
High	No	In the Channel	High
High, Chronic	No	In the Path	High

Table 3.3. Determining Potential for Debris Accumulation on a Single Pier.

Accumulation Potential	Pier type	Location Category	Potential for Debris Delivery
Low	-	Sheltered	-
	Solid Pier	Bank/Floodplain	-
	Solid Pier	In the Channel	Low
	Piers w/ Openings	Bank/Floodplain	Low
Medium	Solid Pier	In the Channel	High
	Solid Pier	In the Path	Low
	Piers w/ Openings	Bank/Floodplain	High
	Piers w/ Openings	In the Channel	Low
High	Solid Pier	In the Path	High
	Piers w/ Openings	In the Path	Low
	Piers w/ Openings	In the Channel	High
High, Chronic	Piers w/ Openings	In the Path	High

Both of these tables were generated from the information presented by Diehl.[17] As shown in these tables, the potential for debris accumulation is based on the estimated delivery potential, which is the same for the entire site, the location category, and the effective length of the span between fixed elements relative to design log length (gap wider or narrower) for span accumulations and the presence or absence of narrow openings that carry flow for single pier accumulations.

3.5   SIZE OF DEBRIS ACCUMULATION AT STRUCTURES

The last phase involves calculating the hypothetical accumulation over the entire length of the bridge with a medium, high, and high, chronic potential. The hypothetical accumulation with a medium and high potential should be used to evaluate the effects that the accumulation would have on the hydraulic characteristics through the bridge and on the hydraulic loading on the structure. The hypothetical accumulation with a high, chronic potential can be used to define the potential maintenance requirements for the bridge, i.e., the location and maximum extent of debris removal.

The potential for debris accumulation estimated in the preceding section is related to the likelihood of occurrence relative to the various components of the bridge, and it does address the likely size of an accumulation. A pier with a "high potential" for accumulation indicates that there is a high potential for accumulation at the pier relative to the potential for accumulation at the other piers, and an accumulation will not necessarily form on such a pier. If an accumulation does form, it may be as wide as the design log length and extend vertically to the depth of flow, or it may be much smaller. The size of the accumulations depends mostly on the debris dimensions and delivery rates, the flow depth, and the number and proximity of gaps and piers affected. Accumulations in the channel can reach their maximum size during a single flood where delivery is high, but accumulations grow more slowly where the debris delivery is low or when the accumulation is outside of the channel.

Diehl[17] proposed that accumulations on a single pier should have a width equal to the design log length over its full flow depth, accumulations across two or more piers should extend laterally half of the design log length beyond them, and accumulations on vertical and horizontal gaps should extend across the entire width and height of the gap. Diehl based this proposal upon conservative assumptions consistent with his largest observed debris accumulations.[17]

Because of limited descriptions and observations available for debris accumulations on superstructures, Diehl could not provide a means to estimate the maximum size of accumulations on superstructures.[17] Consequently he recommended following the suggestions provided by Wellwood and Fenwick[64], which define the vertical extent of the accumulation being 1.2 m (4 ft) above the top of the bridge parapet wall and below the low chord elevation. Based on the above information, the maximum extent of debris accumulations is summarized in Table 3.4.

Table 3.4. Maximum Extent of Debris Accumulation.

Accumulation Type	Width	Height
Pier	Design Log Length	Flow Depth
Superstructure	Span Width	Vertical Height of Superstructure plus 1.2 m Above and Below the Superstructure
Horizontal Gap	Width of Gap	Smaller of Vertical Height of Gap or Flow Depth
Vertical Gap	Width of Gap	Vertical Height of Gap

The overall potential for debris accumulation at a bridge should be defined by the highest potential estimated for the different bridge components. Therefore, a bridge should be considered as a high potential for accumulation if any of its components have been determined to have a high potential for accumulation.

Debris accumulations over the entire length of the bridge should be developed and evaluated for both medium and high potential conditions, with the second condition reflecting a condition that is more likely to occur than the first condition. As proposed by Diehl[17], an accumulation over the entire length of the bridge with a medium potential can be represented by assuming that all of the individual medium- and high-potential accumulations grow to their maximum size. Similarly, an accumulation with a high potential can be defined by assuming that all of the individual high-potential accumulations grow to their maximum size.

As previously mentioned, debris accumulations can cause significant changes in the hydraulic characteristics through the bridge and the trapping characteristics at the bridge. These changes could cause an increase in the potential for accumulation at the bridge. For example, a high-potential blockage across the channel may cause skewed flow through the bents that were initially considered to be aligned with the flow, or the superstructure could become immersed as a result of the increased backwater caused by the debris accumulation on the structure. Therefore after the initial assessment, the bridge should be re-evaluated using the bridge comprised of the debris accumulation determined from the initial assessment and the corresponding hydraulics associated with the accumulation.

Finally, the overall potential for debris accumulation at bridges depends on the probability or frequency of occurrence for the events that were used to define the potential for debris delivery and accumulation at the bridge. A high-potential assessment based on a large flood event and significant changes in the watershed and upstream channel would have different implications on the bridge design and maintenance compared to a high-potential assessment based on a 2-year flood event and existing channel conditions.

3.6   FACTORS THAT AFFECT DEBRIS PRODUCTION

There are several different factors that can influence debris production. These factors include floods, fires, urbanization, logging, land clearing (i.e., grazing and agriculture), conservation practices, and channel improvements.

Flooding increases debris production as the associated discharges serve as a means to produce and deliver debris to a site. Higher discharges are more likely to cause erosive forces on bank and floodplains. The inundation of the flooding event affects more of the floodplain area; facilitating transport of debris into the main channel.

Fires can decrease the amount of floating debris introduced into the stream system. However, fires increase the magnitude of runoff from the burned area, increase the erodibility of soils, and increase the probability of catastrophic events such as debris flows and landslides, resulting in a significant increase in sediment yield from the effected area. This increase could cause an increase in fine and coarse detritus to be transported to and deposited at a culvert or bridge structure.

Urbanization over time causes an opposite effect in the yield of sediment from a watershed than that of fires. Initially, sediment yield can be significantly increased during the construction phase of development due to the removal of exiting vegetation and disturbance of the soil. However over time, the sediment yield decreases as the developed land becomes restabilized and land surface area exposed to the erosive effects of rainfall and runoff is reduced as a result of the increase in impervious area, such as roads, structures, and parking lots. Hydrological effects from urbanization include an increase in runoff volume, higher peak flows, and longer durations. These effects with the decrease in the sediment yield from the watershed could result in an increase in bank erosion and scour of the streambed, which could increase the generation and delivery of floating debris to a bridge site.

Logging has been identified as a source of floating debris.[14,23,25] A study conducted by Froehlich[25] indicated that different logging practices cause substantial differences in the loads of floating debris. Practices that reduce the quantities of floating debris include directional felling uphill with a tree-pulling system and providing a buffer strip of undisturbed vegetation along the streams.

Land Clearing associated with logging, grazing, or agriculture practices could cause the same effects associated with fires, however the magnitude of these effects would most likely not be as severe. Also, grazing allowed near a stream can result in a significant increase in bank erosion.

Conservation practices have the opposite effects than the effects associated with clearing of the land. Implementation of a different conservation practice can reduce both the amount of erosion and runoff from the land.

Channel improvements or modifications to the channel geometry and/or vegetation clearing from the channel can influence quantities of both floating debris and fine/coarse sediment. Improper design of such improvements can cause significant instabilities to develop within the system, including increased bank erosion, increased degradation and/or aggradation of the streambed, and/or significant changes in the planform, that could increase the generation and delivery of floating debris to a structure site.

Growth of riparian forest buffer strips has been recommended and encouraged by the US Environmental Protection Agency (US EPA) for their water quality, ecological and bank erosion benefits especially in agricultural areas. These forested buffer strips adjacent to stream channels are now common with maturing trees especially in heavily agricultural areas.

Extreme events, such as ice storms, debris flows, forest fires and insect infestations can drastically increase the debris load at some point in the life of the structure.

(page intentionally left blank)

CHAPTER 4 – ANALYZING AND MODELING DEBRIS IMPACTS TO STRUCTURES

4.1   INTRODUCTION

After the location and extent of the debris accumulation on the bridge structure has been determined using the procedures discussed in the previous Chapter, a hydraulic analysis should be conducted to evaluate the affects the accumulation would have on the hydraulic characteristics through and upstream of the bridge structure, local scour at the piers, and hydraulic loading on the structure. General information for performing such an analysis is presented in this Chapter. This Chapter also includes information for estimating local pier scour and hydraulic loading on the bridge structure associated with debris accumulation.

4.2   HYDRAULIC ANALYSES OF DEBRIS

Hydraulic analyses of affects of debris upon a drainage structure are often conducted using a one-dimensional (1-D) water surface model. However, such analyses can also be performed using:

- Hand calculations;

- Two-dimensional (2-D) numerical (computer) models;

- Three-dimensional (3-D) computer models; or

- Physical (laboratory) modeling.

The selection of any such analytical technique is based on the complexity of the hydraulics and debris, risk and importance of the drainage structure, and other project site characteristics.

4.3   ONE-DIMENSIONAL DEBRIS ANALYSIS MODELING

One-dimensional programs available for performing debris analyses include the U.S. Army Corps of Engineers (USACE) River Analysis System (HEC-RAS) and the FHWA Water Surface Profile (WSPRO) (among others). For culverts, the FHWA HY-8 program allows evaluation of complex hydraulic conditions.

4.3.1   Data Requirements

Data required for the hydraulic analyses include geometric and flow data. The geometric data consists of cross section data, reach length, energy loss coefficients, and hydraulic structure data. Flow data includes the discharges used in the analyses and the associated boundary conditions and regimes. A thorough discussion of this information, as well as modeling approaches is beyond the scope of this document. However, such information can be found in appropriate user manuals and model documentation. A brief description of each of these modeling data follows:

Cross Section Geometry is the representation of the ground surface perpendicular to the direction of flow. Cross sections are located along a watercourse to define the conveyance capacity of the main channel and the adjacent floodplain. Cross sections are required at representative locations throughout the watercourse and at distinct locations where changes occur in discharge slope, shape, or roughness, or at location where hydraulic structures are located. In modeling debris, the cross section data can be modified to include or simulate ineffective flow areas, levees, and/or blocked obstructions.

Reach Length is the measured distance between cross sections. Reach lengths are provided for the main channel, measured along the thalweg, and for the left and right overbanks, measured along the anticipated path of the center of mass of the overbank flow. In debris analyses, reach lengths serve to allow refined characterization of the extent of the debris field and the associated effects.

Energy Loss Coefficients estimate losses caused by the resistance to flow from bed-surface and vegetative roughness (i.e., Manning's n coefficient) [8,15,28,3,15], channel irregularities, channel alignment, obstructions, and by the contraction and expansion of the flow. Adjusting these coefficients allows simulation of the presence and extent of debris. These also provide a means to simulate the impacts and effects of debris.

Hydraulic Structure Data is the geometric representation of structures that influence the water surface profile within a watercourse. The hydraulic structures can include bridges, culverts, spillways, diversion structures, weirs, etc. The information required to define the bridge structure are the dimensions of the bridge deck, piers, and bridge abutments. The geometry of the debris accumulation should also be accounted for when defining these features. Typically, the dimensions of these features have to be manually adjusted to account for the debris accumulation.

Flow Data required for the model is the discharge in the watercourse and the flow conditions at the boundaries of the model. The discharge is based on the peak discharge for the design flood event of the bridge or for a specific flood event that is being used to estimate the hydraulic loads on the bridge structure. For this discharge, the flow depth is required at the downstream boundary for subcritical flow (Froude number less than 1) and at the upstream boundary for supercritical flow (Froude number greater than 1) to initiate the water surface profile computations.

4.3.2   Background of Modeling Methods and Approaches

There are several methods available for evaluating the hydraulics through a bridge structure for a one-dimensional flow analysis. The type of methods available depends on the flow conditions through the bridge. The flow through the bridge could be classified as either low or high flow conditions.

4.3.2.1 Low Flow Conditions

Low flow conditions exist when the flow through the bridge opening is open channel flow, i.e., the water surface is below the highest point on the low chord of the bridge opening. Three types of flow classes can exist for this condition:

1. **Class A** exists when the water surface through the bridge is completely subcritical, i.e., above critical depth;

2. **Class B** exists when the water surface profile passes through critical depth within the bridge structure, which can occur for either supercritical or subcritical flow; and

3. **Class C** exists when the water surface profile through the bridge structure is completely supercritical, i.e., below critical depth.

There are three methods commonly available for computing the hydraulics through the bridge for low flow conditions. These methods are:

1. **Energy Equation** – This method uses the conservation of energy to determine water surface elevations, velocities, and losses in a waterway. This method is best used when the bridge piers are a small obstruction to the flow and the friction losses are the predominate consideration. This method can be used for both supercritical and subcritical flow (Class A, B, and C).

2. **Momentum Equation** –The momentum equation uses the second law of thermodynamics to describe how change of momentum per unit of time in the body of water in a flowing channel is equal to the resultant of all the external forces that are acting on the body. Unlike the energy equation, this method does not account for non-uniform velocity distributions. The momentum equation method is best used when the bridge piers are the dominant contributor to energy losses or when the pier losses and friction losses are both predominant. As in the energy method, this method can be used for both supercritical and subcritical flow (Class A, B, and C). *FHWA does not recommend this method for 1-D bridge hydraulics.*

3. **Yarnell Equation** – The Yarnell equation empirically predicts the change in water surface from just downstream of the bridge to just upstream of the bridge. The equation is based on about 2,600 lab experiments in which the researchers varied the shape of the piers, the width, the length, the angle, and the flow rate.[66] This method is most applicable when the piers are the dominant contributor to energy losses and the flow through the bridge remains subcritical (Class A). *FHWA does not recommend this method for 1-D bridge hydraulics.*

The energy and momentum equation methods can be used for all of the classes for the low flow condition (Class A, B, and C), while the Yarnell equation method is intended only for Class A low flow.

4.3.2.2 High Flow Conditions

High flow conditions exist when the flow through the bridge opening comes in contact with the maximum low chord of the bridge deck. The type of flow conditions that can occur for high flow include pressure flow, a combination of pressure and weir flow, and a combination of weir flow and open channel flow through the bridge. Generally, three computational approaches exist to evaluate high flow conditions.

1. **Pressure Flow Condition** – Pressure flow occurs when the flow comes into contact with the low chord of the bridge deck. The backwater upstream of the bridge associated with this type of flow condition causes the flow through the structure to behave as orifice flow. In general, there are two types of pressure flow that can exist (details of which are described in other FHWA documents).[10] Depending on the conditions, pressure flow can describe a bridge with full submergence of the low chord at the upstream side and open channel flow at the downstream side of the bridge (acting as a sluice gate – Figure 4.1). The second type of pressure flow exists when both the upstream and downstream sides of the bridge are submerged and uses the standard full flowing orifice equation (Figure 4.2).

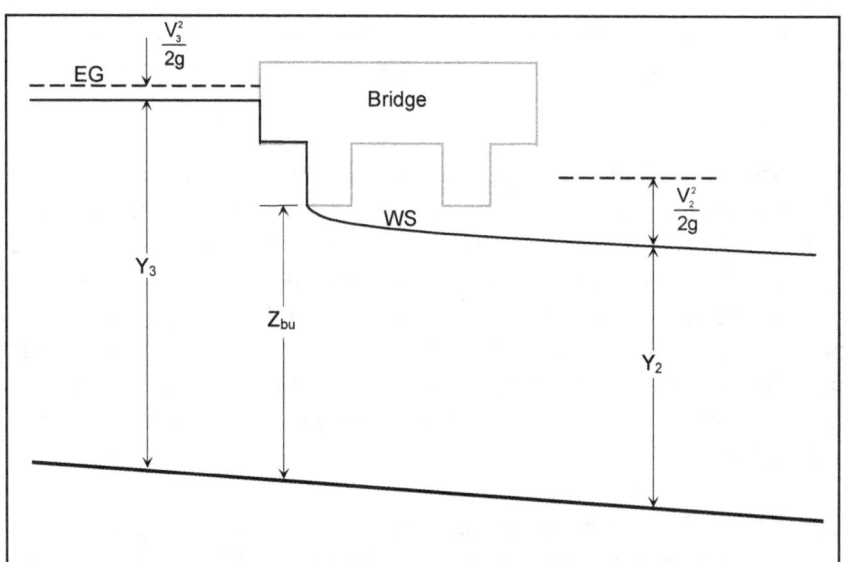

Figure 4.1. Sketch of the sluice gate type of pressure flow.

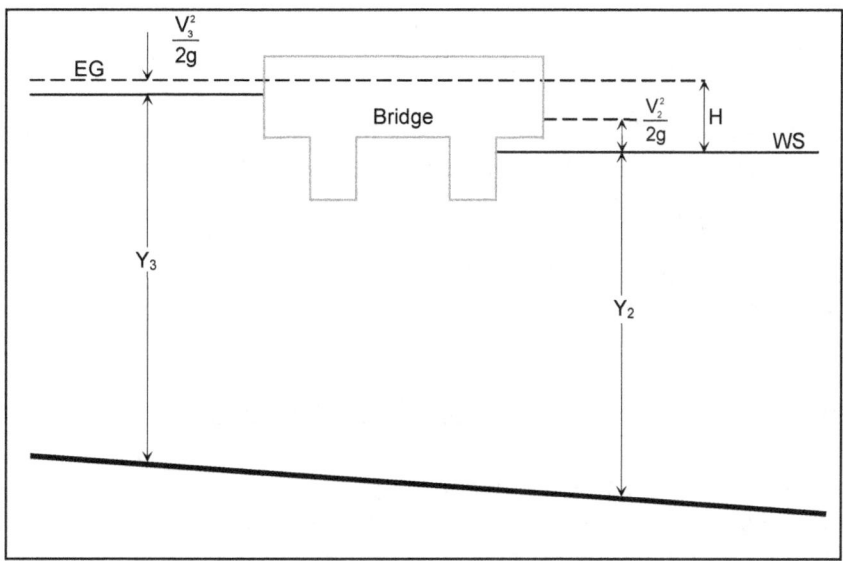

Figure 4.2. Sketch of fully submerged pressure flow.

2. **Weir Flow Condition** – Weir flow exists when water flows over the bridge structure and/or the roadway approaches for the bridge. Typically this condition is modeled using the standard weir equation (Figure 4.3). This illustration depicts pressure flow occurring through the bridge structure, which might not always be the case. Note that pressure flow may also occur through the bridge opening. Typically, some balancing of flow through the opening and over the "roadway" weir occurs before the model converges to a solution.

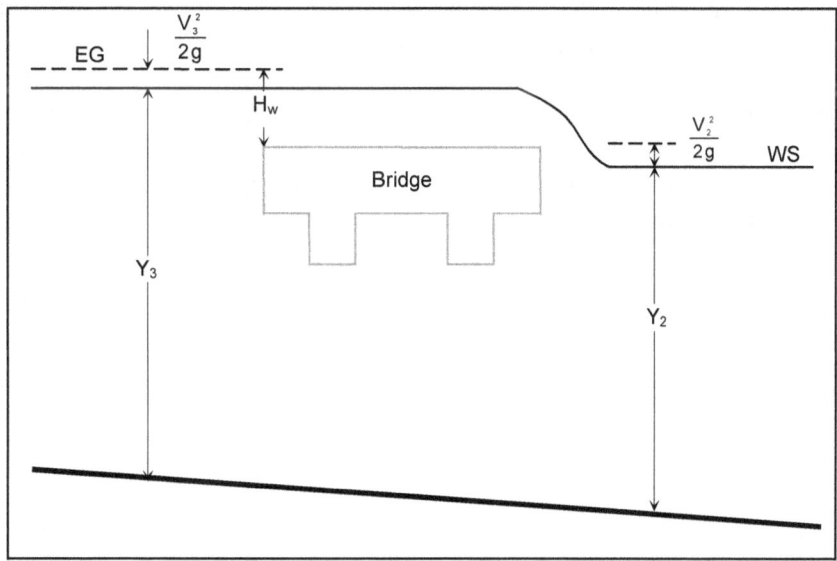

Figure 4.3. Sketch of pressure and weir flow.

3. **Energy Equation Approach** – as with low-flow conditions, the energy method balances the total energy from the downstream side to the upstream side of the bridge structure. All of the computations are performed as though the flow is open channel flow, and the area obstructed by the bridge structure is subtracted from the flow area and the wetted perimeter is increased for the portions of the structure in contact with the water. This method should be used when the bridge is highly submerged and the flow over the road is not controlled by weir flow (low water bridges) or when the bridge deck is a small obstruction to the flow and the bridge opening is not acting like a pressurized orifice. This method is also best used for bridges that are perched above the floodplain.

4.3.3 Scenarios for Hydraulic Modeling of Debris Accumulation

These 1-D model data, methods, and approaches permit computation of water surface profiles at a hydraulic structure with debris accumulation. As previously mentioned, the general concepts discussed would apply to most 1-D models commonly used in the highway hydraulics community.

4.3.3.1 Bridge Debris Scenarios

Scenarios for analyzing debris accumulation at a bridge structure involve relocating cross sections, redefining the ineffective flow boundaries, modifying the cross section and bridge geometry, and changing the contraction and expansion coefficients.

Scenario 1: Relocation of Downstream Wake. As depicted in Figure 4.4, relocating an "expansion" cross section further downstream would attempt to simulate an ineffective flow zone (downstream wake) created by the debris accumulation. The ineffective flow created by the debris accumulation should extend downstream from the upstream face. Assuming a 2:1 to 4:1 expansion ratio for the reach downstream of the bridge, this distance would range from the width (2:1) to twice the width (4:1) of the debris accumulation. This scenario might be required where most of the flow downstream of the bridge is conveyed within the main channel, the overbank areas are not extremely wide, or the bridge structure and/or roadway do not significantly constrict the flow.

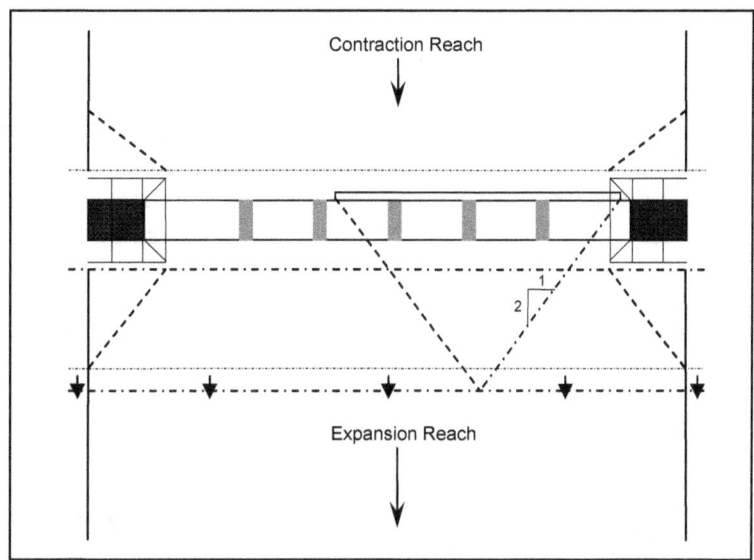

Figure 4.4. Changing downstream expansion cross section location.

Scenario 2: Creating additional downstream ineffective flow boundaries. As seen in Figure 4.5, adding ineffective flow boundaries to downstream bridge face cross section would simulate the *downstream wake* created by the debris accumulation. Once again, assume a 2:1 to 4:1 expansion ratio to create the locations of the ineffective flow. Adding additional downstream cross sections would assist in the transition. This scenario could be used alone or in conjunction with other scenarios.

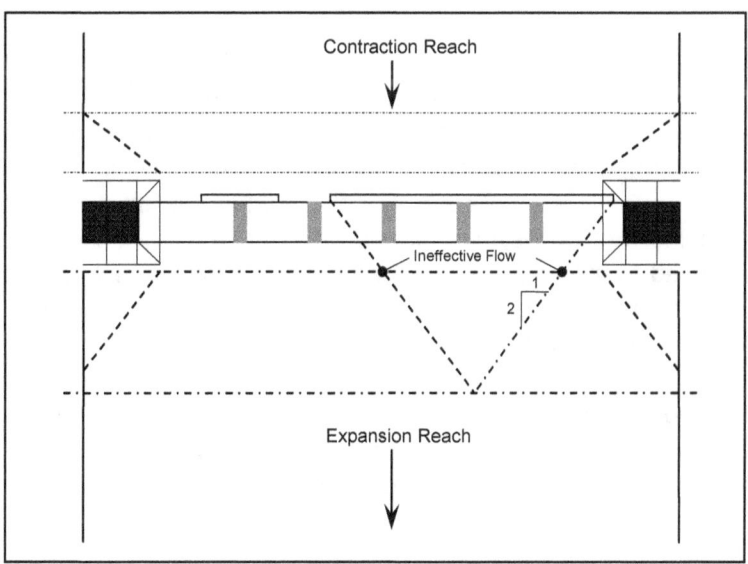

Figure 4.5. Adding downstream ineffective flow locations.

Scenario 3: Creating upstream ineffective flow boundaries. An additional upstream cross section would simulate an ineffective flow zone created by the *debris accumulation* (Figure 4.6). Assuming a 1:1 to 2:1 contraction ratio the reach upstream of the bridge to the debris accumulation, the point where flow is not affected by the accumulation would be located upstream about one-half (1:1) to the entire (2:1) of the debris accumulation width. Depending on the accumulation width, adding additional upstream cross sections would assist in the transition. This scenario could be used alone or in conjunction with other scenarios.

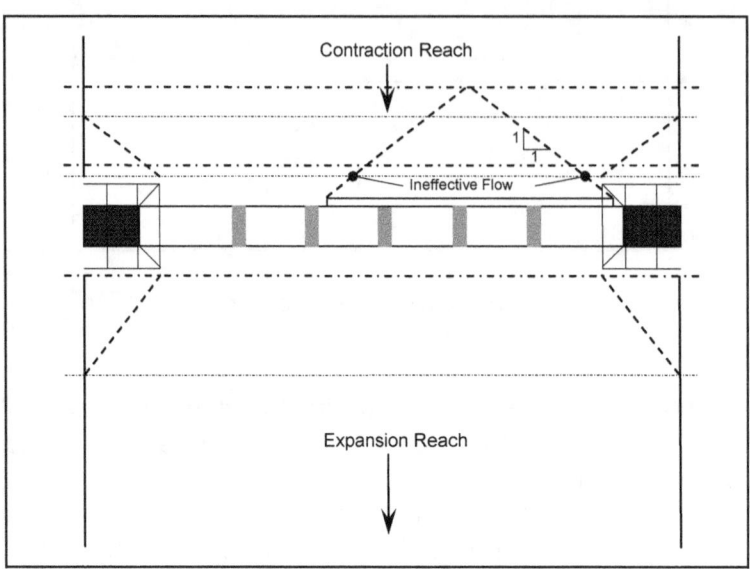

Figure 4.6. Adding upstream ineffective flow locations.

Scenario 4: Modifying Bridge Geometry. Modification of the bridge geometry could be used to reflect debris accumulation. The modification could range from changing local structural elements, such as piers and abutments, to actual changes in the bridge opening area or low chord elevations. For piers, debris accumulation would change how pier width and height would be described in model input. The debris may collect either *symmetrically* or *asymmetrically* about the pier centerline. For an asymmetrical debris accumulation, the centerline of the pier would have to be moved to the centerline of the debris accumulation or an additional "dummy" bridge pier with an extremely narrow width would have to be defined at the centerline of the debris accumulation. This is illustrated in Figure 4.7.

Scenario 5: Modifying Contraction and Expansion Losses. In some cases, increasing the contraction and expansion loss coefficients may be appropriate if the debris accumulation causes an abrupt contraction and expansion, respectively, of the flow.

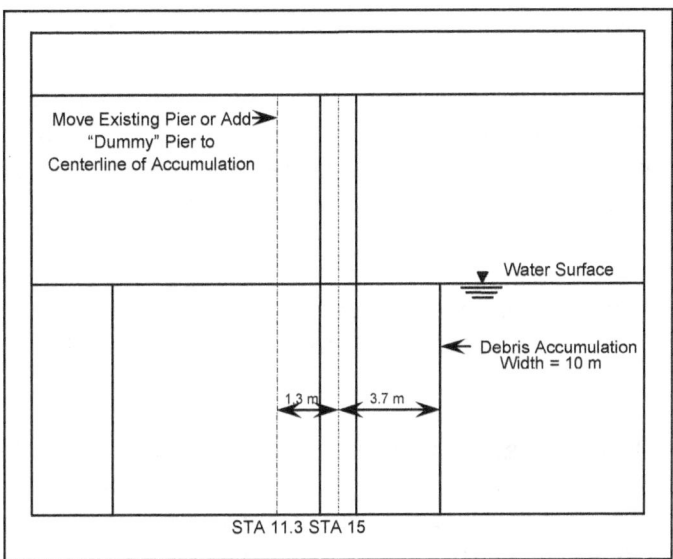

Figure 4.7. "Dummy" bridge pier used to simulate an asymmetrical debris accumulation.

The results of the hydraulic analysis can be used to estimate the hydraulic loading on the bridge structure associated with the debris accumulation. Information required from a 1-D model to estimate the loads is briefly discussed in the following paragraphs.

Upstream Water Surface Elevation. The upstream water surface elevation is used to compute the blockage ratio, B, the hydrostatic and hydrodynamic forces at the upstream side of bridge, and the location of the forces. This elevation should be selected at a location upstream of where the flow accelerates in response to the debris accumulation, causing flow-separation and a zone of ineffective flow. The upstream boundary of this zone is depicted in Figure 4.6. As shown in this figure, the boundary is based on the flow contraction ratio typically assumed near a bridge structure. Based on this assumption, the upstream water surface elevation should be obtained from a cross section located upstream of the debris accumulation at a minimum distance of at least one half of the total width of the debris accumulation.

Downstream Water Surface Elevation. The downstream water surface elevation is used to compute the downstream hydrostatic forces on the bridge structure that is used to determine the net hydrostatic forces on the structure. As in the upstream water surface elevation, this elevation should be obtained sufficiently far downstream from the structure that the flow is not affected by the wake (ineffective flow zone) created by the debris accumulation (see Figure 4.5). Based on the assumptions discussed for the upstream water surface elevation, the downstream water surface elevation should be obtained from a cross section located downstream of the debris accumulation at a distance equal to the total width of the debris accumulation. The typical response within this reach is for the water surface elevation to recover from the drop in the water surface elevation (increased velocities) through the contracted section by increasing downstream of the bridge. However, there are some conditions (high rates of energy dissipation, large channel slopes, or large changes in channel geometry) where the water surface elevation downstream does not recover (rise in elevation) from the flow contraction. For

4.9

these conditions, the water surface elevation downstream from the effects of the debris may be lower than the elevations in the contraction of the bridge section. For such cases, the water surface elevation in the contracted section should be used to compute the downstream hydrostatic forces. To summarize, the water surface elevation used to compute the downstream hydrostatic forces should be based on the higher water surface elevation at either the contracted section or the cross section immediately downstream of the wake created by the debris accumulation.

Area within the Contracted Section. The flow area within the contracted section is used to compute the blockage ratio, B. The area should be based on the smallest area within the bridge section.

Flow Velocity within the Contracted Section. The flow velocity within the contracted section is used to compute the dynamic forces on the structure. For accumulations on piers, the reference location for the velocity depends on the percentage of blockage of the debris and bridge structure. If the reduction in area is anticipated to be greater than 30 percent of the entire wetted cross-sectional flow area in the bridge opening, then the reference velocity is based on the maximum average velocity in the contracted section of the entire bridge opening. If the reduction is anticipated to be less than 30 percent, then the reference velocity is based on the maximum local average velocity near the pier and debris accumulation, i.e., maximum average flow velocity in the main channel for piers located in the main channel and maximum average flow velocity in the left overbank for piers located in the left overbank. For accumulations on superstructures, the reference velocity is the maximum contracted flow velocity in main channel under the superstructure for any degree of blockage.

Average Flow Depth within the Contracted Section. The average flow depth within the contracted section is used to compute the Froude number that is utilized in selecting the drag coefficient. This depth should be based on the same area used to define the reference velocity, i.e., the average depth in the main channel should be used if the reference velocity is based on the average flow velocity in the main channel.

4.3.3.2   Culvert Debris Scenarios

Most of the common 1-D models have some culvert hydraulic capabilities incorporated within their algorithms. In several cases, these capabilities would allow application of the bridge-based scenarios to these culvert structures. However, these 1-D models do not simulate every barrel type or configuration. Additionally, for certain hydraulic and discharge conditions, these 1-D models may not replicate underlying assumptions of culvert hydraulics.

Additionally, there is not a great deal of research available into the effects of debris upon hydraulic performance of culverts. Therefore, FHWA recommends use of specific culvert models, specifically HY-8, for both culvert hydraulic and debris analyses. In such use, debris analysis scenarios would modify barrel parameters to reflect changes in inlet efficiency (i.e., entrance loss coefficients), additional roughness, or reduced equivalent barrel area.

4.4   ADVANCED MODELING

Flow, geometric, boundary, topographic, or other conditions near the bridge structure may necessitate use of 2-D numeric, 3-D numeric, or physical models. Such models allow a more detailed prediction of the flow-separation regions and hydraulic pressure variations near the bridge structure and debris accumulation than what is assumed in the one-dimensional analysis.

These models can also be very useful in defining the locations of the stream channel where high-debris transport would most likely occur. The numerical (i.e., 2-D and 3-D) models need approximately the same type of information as the 1-D model, however they use different means to represent data including:

1. Using a finite element mesh of nodes and links to represent the topography, bathometry, and drainage structures;

2. Requiring the resistance coefficient and turbulence parameter be defined at each node; and

3. Defining the boundary conditions differently.

The physical models require scaled design of the site to allow predictions of flow and debris characteristics and tendencies.

Use of such models would be predicated on the relative importance of simulating the debris accumulation for a project. Several research projects are investigating the appropriate use of such advanced models in situations such as these. Additionally, increased power of computers and hydraulic software packages are making such analyses more cost effective for the transportation community.

4.5   LOCAL PIER SCOUR ASSOCIATED WITH DEBRIS ACCUMULATION

Debris accumulations on a bridge pier can increase local scour at the pier as a result of increased pier width and downward flow component upstream of the pier. When debris accumulates on a pier, the scour depth can be estimated by assuming that the pier width is larger than the actual width. A width equal to the design log length as defined by Diehl[17] and presented in the previous chapter of this manual can be assumed for estimating the scour at the pier. This assumption could be on the conservative side at large depths because the effect of the debris on scour depth diminishes.

Only limited research exists on local scour at piers with debris accumulation. Melville and Dongol have conducted a limited quantitative study of the effect of debris accumulation on local pier scour and have made some recommendations which support the approach suggested above[39]. An interim procedure for estimating the effect of debris accumulation on local scour at piers is presented in Appendix D of HEC-18[55].

4.6   HYDRAULIC LOADING ASSOCIATED WITH DEBRIS ACCUMULATION

There are three steps for computing the hydraulic loading on a bridge structure with debris accumulation. The first step is to define the geometry of the debris accumulation using the procedures and recommendations presented in Chapter 3 of this manual. The second step is to compute the flow hydraulics through the bridge structure using the procedures and recommendations presented in the previous sections of this chapter. The last step is to compute the hydrodynamic loads using the hydraulic characteristics associated with the presence of the debris accumulation and the following equations and general procedure developed by Parola[49].

The hydrodynamic drag force is based on the general form of the drag equation and the drag coefficient relationship developed from a model study investigation by Parola at the University of Louisville[49].

$$F_D = C_D \gamma A_D \frac{V_r^2}{2g} \tag{4.1}$$

where:

F_D	=	Drag force, N (lbs)
C_D	=	Drag coefficient, see Tables 4.1 and 4.2
γ	=	Specific weight of water, N/m^3 (lbs/ft^3)
A_D	=	Area of wetted debris based on the upstream water surface elevation projected normal to the flow direction, m^2 (ft^2)
V_r	=	Reference velocity, see discussion in Subsection 4.3.3.1, m/s (ft/s)
g	=	Acceleration of gravity, 9.81 m/s^2 (32.2 ft/s^2)

Drag coefficient for debris on piers is provided in Table 4.1 and for debris on superstructures in Table 4.2.

Table 4.1. Drag Coefficient for Debris on Piers.

Value of B	Value of F_r	C_D
B < 0.36	F_r < 0.4	1.8
B < 0.36	0.4 < F_r < 0.8	2.6 – 2.0F_r
0.36 < B < 0.77	F_r < 1	3.1 – 3.6B
B > 0.77	F_r < 1	1.4 -1.4B

Table 4.2. Drag Coefficient for Debris on Superstructure.

Value of B	Value of F_r	C_D
B < 0.33	F_r < 0.4	1.9
B < 0.33	0.4 < F_r < 0.8	2.8 – 2.25F_r
0.33 < B < 0.77	F_r < 1	3.1 – 3.6B
B > 0.77	F_r < 1	1.4 -1.4B

The drag coefficient as provided in these tables is related to the blockage ratio and Froude number as defined below.

$$B = \frac{A_d}{A_d + A_c}$$ (4.2)

where:

B	=	Blockage ratio
A_d	=	Cross-sectional flow area blocked by debris in the contracted bridge section, m² (ft²)
A_c	=	Unobstructed cross-sectional flow in the contracted section, m² (ft²)

$$F_r = \frac{V_r}{\sqrt{gy_r}}$$ (4.3)

where:

Fr	=	Froude number
V_r	=	Reference velocity, see discussion in Subsection 4.3.3.1, m/s (ft/s)
g	=	Acceleration of gravity, 9.81 m/s² (32.2 ft/s²)
y_r	=	Average flow depth corresponding with the reference velocity, m (ft)

The total force on the structure that is caused by the hydrostatic pressure difference can be approximated as:

$$F_h = \gamma(h_{cu}A_{hu} - h_{cd}A_{hd})$$ (4.4)

where:

F_h	=	Horizontal hydrostatic force on area A_h, N (lbs)
γ	=	Specific weight of water, N/m³ (lbs/ft³)
h_{cu}	=	Vertical distance from the upstream water surface to the centroid of area A_{hu}, m (ft)
A_{hu}	=	Area of the vertically projected, submerged portion of the debris accumulation below the upstream water surface, m² (ft²)
h_{cd}	=	Vertical distance from the downstream water surface to the centroid of area A_{hd}, m (ft)
A_{hd}	=	Area of the vertically projected, submerged portion of the debris accumulation below the downstream water surface, m² (ft²)

The total resultant force is computed as the summation of the drag force (Equation 4.1) and the differential hydrostatic force (Equation 4.4). The loads computed using these equations corresponds to the pressure forces of the water on the debris accumulation. The transfer of the load from the debris to the structure depends on many factors, including the characteristics of the debris accumulation and the degree to which streambed and banks support the debris accumulation. Thus, a conservative approach of applying the resultant force as a point load is recommended in evaluating the forces on the structure. The vertical and horizontal location of the resultant hydrostatic and drag forces and that of the total force can be determined by adding the moments about convenient axes. A less conservative distribution of the load to the structure may be warranted where there is more information available on the debris configuration and structural susceptibility.

Three scenarios should be evaluated when debris accumulation exists on two piers as a result of the opening between the piers being less than the length of the design log:

1. Debris accumulation of maximum effective width (design log length) forms on Pile Bent 1, with a smaller effective accumulation on Pile Bent 2;

2. Debris accumulation of maximum effective width forms on Pile Bent 2, with a smaller effective accumulation on Pile Bent 1; and

3. A large log spans the opening and transfers or divides the load on the accumulation between the piers almost equally to each pier. Although the pressures on the debris accumulation are almost identical for each scenario, the distribution of the total force to each of the piers may be substantially different for each of the scenarios.

Debris accumulations typically align themselves with the direction of the flow. There is a lot of uncertainty associated with debris accumulation geometry and the direction of the flood flows. Therefore, the resultant force should be applied using both consideration of the anticipated range of possible flow directions and the structure's susceptibility to the resultant forces over the range of flow direction. For example, if the possible direction of flow is 20 degrees to the axis of the pier and the pier is most susceptible to a force applied at 15 degrees, then the force should be applied at 15 degrees to the axis of the pier. For superstructures and debris accumulations that span adjacent piers, the forces should be applied in at least two directions: (1) perpendicular to the face of the bridge and (2) in the direction of the flow with consideration to the structures susceptibility.

4.7 HYDRAULIC LOADING EXAMPLE PROBLEMS

4.7.1 <u>Example 1 – Hydraulic Loading on a Single Pier (SI)</u>

Given:

Design flow rate = 195 m³/s

Minimum upstream main channel width = 13.7 m; design log length of 13.7 m

Depth of debris is full-flow depth

Main channel width at the bridge = 60 m

Debris accumulation only on Pile Bent 2 (see Figure 4.8)

Superstructure is not submerged

Ineffective flow areas from the debris defined by 1:1 contraction and 2:1 expansion

Bottom elevation of Pile Bent 2 = 61.78 m

Left station of debris = 160.33 m; Right station of debris = 174.03

Hydraulic computation results are provided in Table 4.3 and shown in Figure 4.9

Upstream water surface elevation, WS_{US} = 65.43 m (Table 4.3)

Downstream water surface elevation, WS_{DS} = 65.06 m (Table 4.3, see discussion <u>Downstream Water Surface Elevation</u> in Subsection 4.3.3.1)

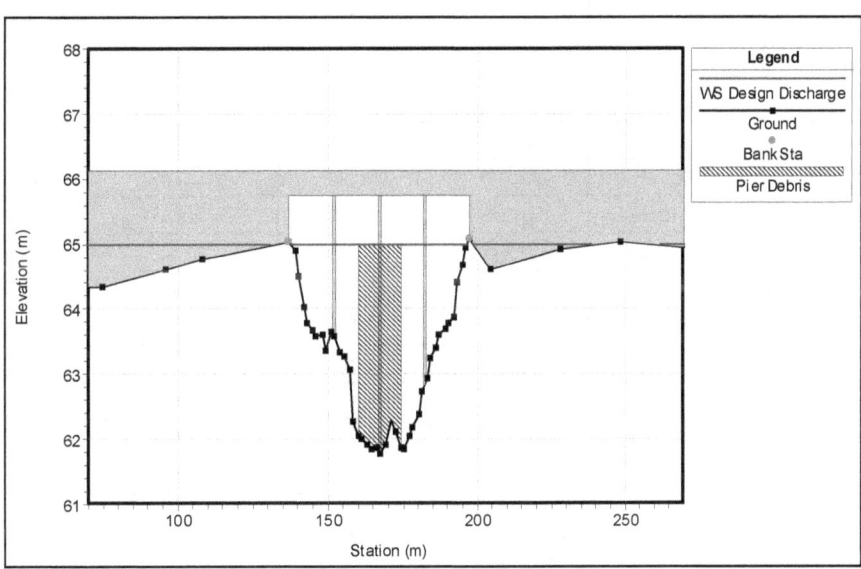

Figure 4.8. Upstream face of the bridge for Example 1.

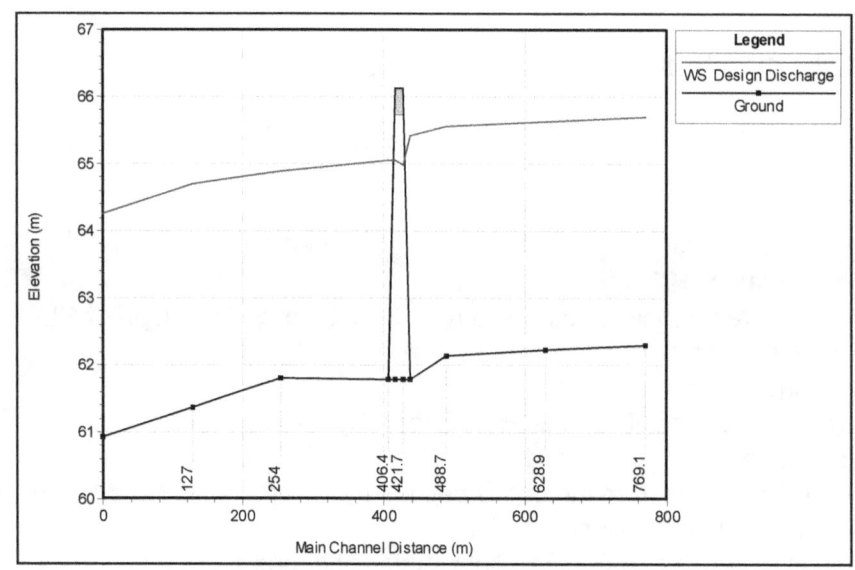

Figure 4.9. Water surface profile for Example 1.

Table 4.3. Results of Hydraulic Calculations for Example 1.

River Station (m)	Water Surface Elevation (m)	Flow Area (m^2)	Main Channel Velocity (m/s)	Cross Section Average Velocity (m/s)	Average Flow Depth[1] (m)
0	64.27	230.98	2.48	0.84	2.34
127	64.70	300.10	1.59	0.65	2.08
254	64.89	307.65	1.03	0.63	1.55
406.4	65.05	123.84	1.66	1.57	1.95
421.7 BR D	65.06	110.19	1.77	1.77	1.94
421.7 BR U	64.99	67.96	2.87	2.87	1.59
436.9	65.43	149.06	1.39	1.31	2.32
488.7	65.56	415.13	1.23	0.47	2.77
628.9	65.63	439.48	1.29	0.44	2.73
769.1	65.71	464.02	1.19	0.42	2.55

Notes:

1. For this example, the entire bridge opening was defined as the main channel. So, the average depth of the main channel is the same as the average depth of the entire cross section.

Determine:

Compute the hydrostatic and drag forces for a debris accumulation on a single pier.

Solution:

Hydrostatic Force on Single Pier Accumulation

A_{hu} = Area of the debris accumulation below the upstream water surface
A_{hu} = (WS$_{US}$ – Debris bottom, DB$_{EL}$)(Width of debris accumulation, W$_D$)
A_{hu} = (65.43 – 61.78)(13.7) = 50.01 m^2

A_{hd} = Area of the debris accumulation below the downstream water surface
A_{hd} = (WS$_{DS}$ – DB$_{EL}$)(W$_D$)
A_{hd} = (65.06 – 61.78)(13.7) = 44.93 m^2

h_{cu} = Vertical distance to centroid of A_{hu} = 0.5(WS$_{US}$ – DB$_{EL}$)
h_{cu} = 0.5(65.43 – 61.78) = 1.83 m

h_{cd} = Vertical distance to centroid of A_{hd} = 0.5(WS$_{DS}$ – DB$_{EL}$)
h_{cd} = 0.5(65.06 – 61.78) = 1.64 m

F_{hu} = Hydrostatic force upstream = $\gamma h_{cu} A_{hu}$
F_{hu} = (9.81)(1.83)(50.01) = 898 kN

F_{hd} = Hydrostatic force downstream = $\gamma h_{cd} A_{hd}$
F_{hd} = (9.81)(1.64)(44.93) = 723 ken

F_h = Total hydrostatic force on Pile Bent 2 = F_{hu} - F_{hd}
F_h = 898 – 723 = 175 kN

Drag Force on Single Pier Accumulation

$$B = \text{Blockage ratio} = \frac{A_d}{A_d + A_c}$$

$$B = \frac{50.01}{50.01 + 67.96} = 0.42$$

B is greater than 0.3, therefore V_r should be based on the average velocity in the contracted section.

V_r = 2.87 m/s (Table 4.3)

$$Fr = \text{Froude number} = \frac{V_r}{\sqrt{g y_r}}$$

$$Fr = \frac{2.87}{\sqrt{(9.81)(1.59)}} = 0.73$$

C_D = Drag coefficient = 3.1 − 3.6B (Table 4.1)
C_D = 3.1 − (3.6)(0.42) = 1.59

$$F_D = \text{Drag force on Pile Bent 2} = C_D \gamma A_{hu} \frac{V_r^2}{2g}$$

$$F_D = (1.59)(9810)(50.01)\frac{(2.87)^2}{2(9.81)} = 327 \text{ kN}$$

Total Force on Single Pier Accumulation

F = Total segment force = $F_h + F_D$
F = 175 + 327 = 502 kN

Location of Forces on Single Pier Accumulation

$$F_{hEL} = \text{Elevation of hydrostatic force} = DB_{EL} + \frac{F_{hu}\left(\frac{WS_{US} - DB_{EL}}{3}\right) - F_{hd}\left(\frac{WS_{DS} - DB_{EL}}{3}\right)}{F_h}$$

$$F_{hEL} = 61.78 + \frac{898\left(\frac{65.43 - 61.78}{3}\right) - 723\left(\frac{65.06 - 61.78}{3}\right)}{175} = 63.51\,\text{m}$$

F_{DEL} = Elevation of drag force = $0.5(WS_{US} + DB_{EL})$
F_{DEL} = 0.5(65.43 + 61.78) = 63.61 m

$$F_{EL} = \text{Elevation of total force} = \frac{(F_D)(F_{DEL}) + (F_h)(F_{hEL})}{F}$$

$$F_{EL} = \frac{(327)(63.61) + (175)(63.51)}{502} = 63.58\,\text{m}$$

F_{hST} = Station of hydrostatic force = 0.5(Left station of debris + right station of debris)
F_{DST} = Station of drag force = F_{hST}
F_{ST} = Station of total force = F_{DST} = F_{hST}
F_{hST} = F_{DST} = F_{ST} = 0.5(160.33 + 174.03) = 167.18 m

4.18

4.7.2 Example 2 – Hydraulic Loading on Two Adjacent Piers, Case 1 (SI)

Given:

Design flow rate = 85 m^3/s

Minimum upstream main channel width = 13.7 m; design log length = 13.7 m

Depth of debris is full-flow depth

Main channel width at the bridge = 60 m

Superstructure is not submerged

Ineffective flow areas from debris defined by 1:1 contraction and 2:1 expansion

Bottom elevation of Pile Bent 2 = 61.98 m; Pile Bent 3 = 61.92 m

Total accumulation width = 25.4 m (defined by assuming that the accumulation extends laterally half the design log length beyond each pier).

Accumulation width on Pile Bent 2 = 13.7 m for Case 1 and 11.7 for Case 2

Accumulation width on Pile Bent 3 = 11.7 m for Case 1 and 13.7 for Case 2

Pile Bent 2, left station of debris = 154.69 m; Right station of debris = 168.39 m

Pile Bent 3, left station of debris = 168.39 m; Right station of debris = 180.09 m

Hydraulic computation results are provided in Table 4.4 and shown in Figure 4.11

Upstream water surface elevation, WS_{US} = 65.28 m (Table 4.4)

Downstream water surface elevation, WS_{DS} = 64.59 m (Table 4.4, see discussion of <u>Downstream Water Surface Elevation</u> in Subsection 4.3.3.1)

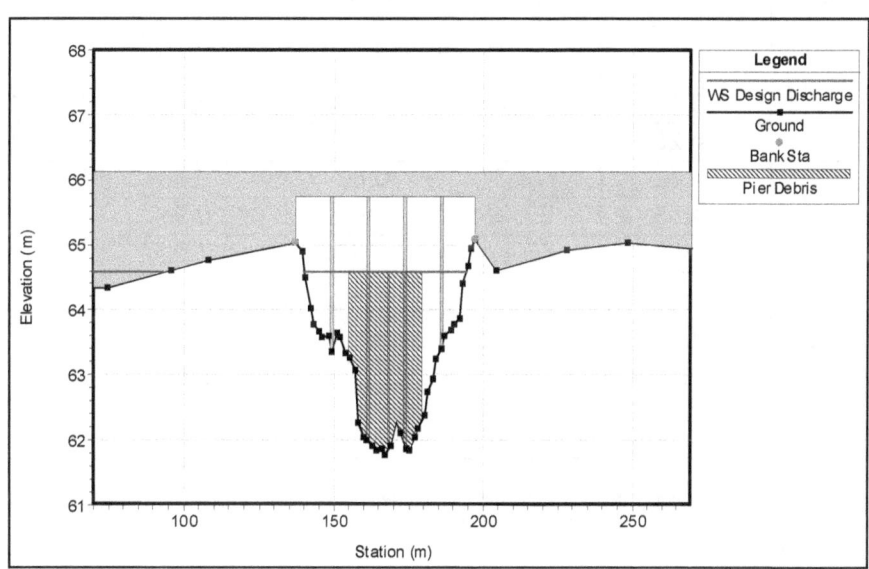

Figure 4.10. Upstream face of the bridge for Example 2.

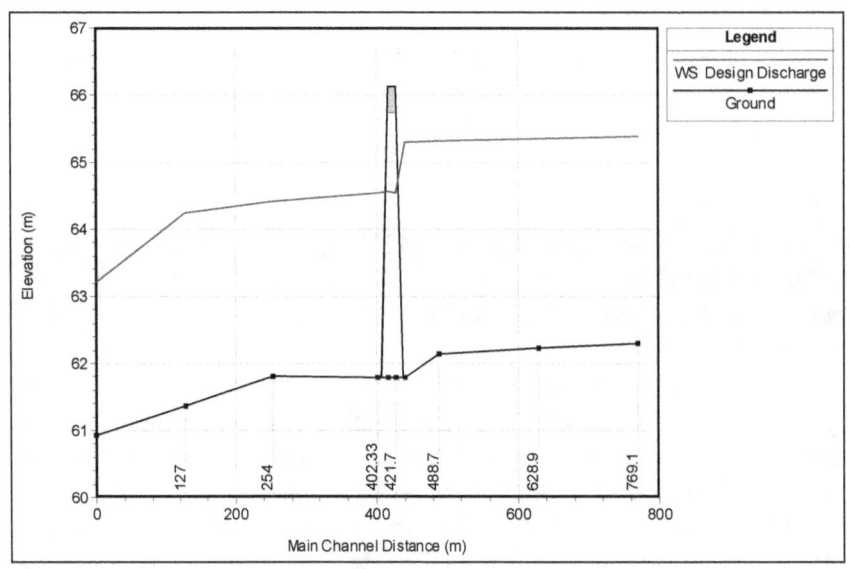

Figure 4.11. Water surface profile for Example 2.

Table 4.4. Results of Hydraulic Calculations for Example 2.

River Station (m)	Water Surface Elevation (m)	Flow Area (m^2)	Main Channel Velocity (m/s)	Cross Section Average Velocity (m/s)	Average Flow Depth[1] (m)
0	63.22	28.81	3.52	2.95	1.38
127	64.25	176.00	1.22	0.48	1.63
254	64.42	176.31	0.82	0.48	1.08
402.33	64.55	90.49	0.96	0.94	1.63
421.7 BR D	64.56	83.47	1.02	1.02	1.63
421.7 BR U	64.54	27.41	3.10	3.10	0.99
440.46	65.30	142.17	0.64	0.60	2.20
488.7	65.32	308.93	0.69	0.28	2.53
628.9	65.35	308.60	0.79	0.28	2.44
769.1	65.38	315.16	0.78	0.27	2.22

Notes:
1. For this example, the entire bridge opening was defined as the main channel. So, the average depth of the main channel is the same as the average depth of the entire cross section.

Determine:

Compute the hydrostatic and drag forces for a debris accumulation on two piers. The calculations are for only Case 1, which is based on the width of the debris accumulation on Pile Bent 2 being equal to the design log length and Pile Bent 3 having a smaller accumulation width. Case 2 is the reverse of Case 1, i.e., the width of the accumulation on Pile Bent 3 would be equal to the design log length and Pile Bent 2 would have a smaller accumulation width. Case 3 is based on the assumption that the design log length spans the opening approximately in the middle of the two piers and the resulting load on the accumulation is transferred equally to each pier.

Solution:

Hydrostatic Forces on Two Adjacent Piers (Case 1)

A_{hu} = Area of the debris accumulation below the upstream water surface
A_{hu} = (WS_{US} – Debris bottom, DB_{EL})(Width of debris accumulation, W_D)
A_{hu} = (65.30 – 61.98)(13.70) = 45.48 m^2 for Pier Bent 2
A_{hu} = (65.30 – 61.92)(11.70) = 39.55 m^2 for Pier Bent 3

A_{hd} = Area of the debris accumulation below the downstream water surface
A_{hd} = (WS_{DS} – DB_{EL})(W_D)
A_{hd} = (64.55 – 61.98)(13.70) = 35.21 m^2 for Pier Bent 2
A_{hd} = (64.55 – 61.92)(11.70) = 30.77 m^2 for Pier Bent 3

h_{cu} = Vertical distance to centroid of A_{hu} = 0.5(WS_{US} – DB_{EL})
h_{cu} = 0.5(65.30 – 61.98) = 1.66 m   for Pier Bent 2
h_{cu} = 0.5(65.30 – 61.92) = 1.69 m   for Pier Bent 3

h_{cd} = Vertical distance to centroid of A_{hd} = 0.5(WS_{DS} – DB_{EL})
h_{cd} = 0.5(64.55 – 61.98) = 1.29 m   for Pier Bent 2
h_{cd} = 0.5(64.55 – 61.92) = 1.32 m   for Pier Bent 3

F_{hu} = Hydrostatic force upstream = $\gamma h_{cu} A_{hu}$
F_{hu} = (9.81)(1.66)(45.48) = 741 kN   for Pier Bent 2
F_{hu} = (9.81)(1.69)(39.55) = 656 kN   for Pier Bent 3

F_{hd} = Hydrostatic force downstream = $\gamma h_{cd} A_{hd}$
F_{hd} = (9.81)(1.29)(35.21) = 446 kN   for Pier Bent 2
F_{hd} = (9.81)(1.32)(30.77) = 399 kN   for Pier Bent 3

F_h = Total hydrostatic force on Pile Bent = F_{hu} - F_{hd}
F_h = 741 – 446 = 295 kN   for Pier Bent 2
F_h = 656 – 399 = 257 kN   for Pier Bent 3

Drag Force on Two Adjacent Piers (Case 1)

$$B = \text{Blockage Ratio} = \frac{A_d}{A_d + A_c} = \frac{A_{hu(pier\,2)} + A_{hu(pier\,3)}}{A_{hu(pier\,2)} + A_{hu(pier\,3)} + A_c}$$

$$B = \frac{45.48 + 39.55}{45.48 + 39.55 + 27.41} = 0.76$$

B is greater than 0.3, therefore V_r should be based on the average velocity in the contracted section.

$V_r = 3.10$ m/s (Table 4.4)

$$Fr = \text{Froude number} = \frac{V_r}{\sqrt{g\,y_r}}$$

$$Fr = \frac{3.10}{\sqrt{(9.81)(0.99)}} = 0.99$$

C_D = Drag coefficient = $3.1 - 3.6B$ (Table 4.1)
$C_D = 3.1 - (3.6)(0.76) = 0.36$

$$F_D = \text{Drag force on Pile Bent} = C_D \gamma A_{hu} \frac{V_r^2}{2g}$$

$$F_D = (0.36)(9810)(45.48)\frac{(3.10)^2}{2(9.81)} = 79 \text{ kN} \quad \text{for Pile Bent 2}$$

$$F_D = (0.36)(9810)(39.55)\frac{(3.10)^2}{2(9.81)} = 68 \text{ kN} \quad \text{for Pile Bent 3}$$

Total Force on Two Adjacent Piers (Case 1)

F = Total segment force = $F_h + F_D$
F = 295 + 79 = 374 kN   for Pile Bent 2
F = 257 + 68 = 325 kN   for Pile Bent 3

Location of Forces on Single Pier Accumulation

$$F_{hEL} = \text{Elevation of hydrostatic force} = DB_{EL} + \frac{F_{hu}\left(\dfrac{WS_{US} - DB_{EL}}{3}\right) - F_{hd}\left(\dfrac{WS_{DS} - DB_{EL}}{3}\right)}{F_h}$$

$$F_{hEL} = 61.98 + \frac{741\left(\dfrac{65.30 - 61.98}{3}\right) - 446\left(\dfrac{64.55 - 61.98}{3}\right)}{295} = 63.46 \text{ m} \quad \text{for Pile Bent 2}$$

$$F_{hEL} = 61.92 + \dfrac{656\left(\dfrac{65.30 - 61.92}{3}\right) - 399\left(\dfrac{64.55 - 61.92}{3}\right)}{257} = 63.43 \text{ m} \quad \text{for Pile Bent 3}$$

F_{DEL} = Elevation of drag force = $0.5(WS_{US} + DB_{EL})$

$F_{DEL} = 0.5(65.30 + 61.98) = 63.64$ m   for Pile Bent 2

$F_{DEL} = 0.5(65.30 + 61.72) = 63.51$ m   for Pile Bent 3

$$F_{EL} = \text{Elevation of total force} = \dfrac{(F_D)(F_{DEL}) + (F_h)(F_{hEL})}{F}$$

$$F_{EL} = \dfrac{(79)(63.64) + (295)(63.46)}{374} = 63.50 \text{ m} \quad \text{for Pile Bent 2}$$

$$F_{EL} = \dfrac{(68)(63.51) + (257)(63.43)}{325} = 63.45 \text{ m} \quad \text{for Pile Bent 3}$$

F_{hST} = Station of hydrostatic force = 0.5(Left station of debris + right station of debris)

F_{DST} = Station of drag force = F_{hST}

F_{ST} = Station of total force = F_{DST} = F_{hST}

$F_{hST} = F_{DST} = F_{ST} = 0.5(154.69 + 168.39) = 161.54$ m   for Pile Bent 2

$F_{hST} = F_{DST} = F_{ST} = 0.5(138.39 + 180.09) = 174.24$ m   for Pile Bent 3

4.7.3   Example 3 – Hydraulic Loading on a Superstructure (SI)

Given:

Design flow rate = 220 m^3/s
Low chord elevation of bridge = 65.5 m
Depth of debris is 1.2 meters below the bridge low chord = 64.3 m
Debris accumulation extends along the entire length of the structure (see Figure 4.12)
Main channel width at the bridge = 60.0 m
Left station of debris = 137.16 m; Right station of debris = 197.21 m
Hydraulic computation results are provided in Table 4.5 and shown in Figure 4.13
Upstream water surface elevation, WS_{US} = 65.71 m (Table 4.5)
Downstream water surface elevation, WS_{DS} = 65.13 m (Table 4.5, see discussion of Downstream Water Surface Elevation in Subsection 4.3.3.1)

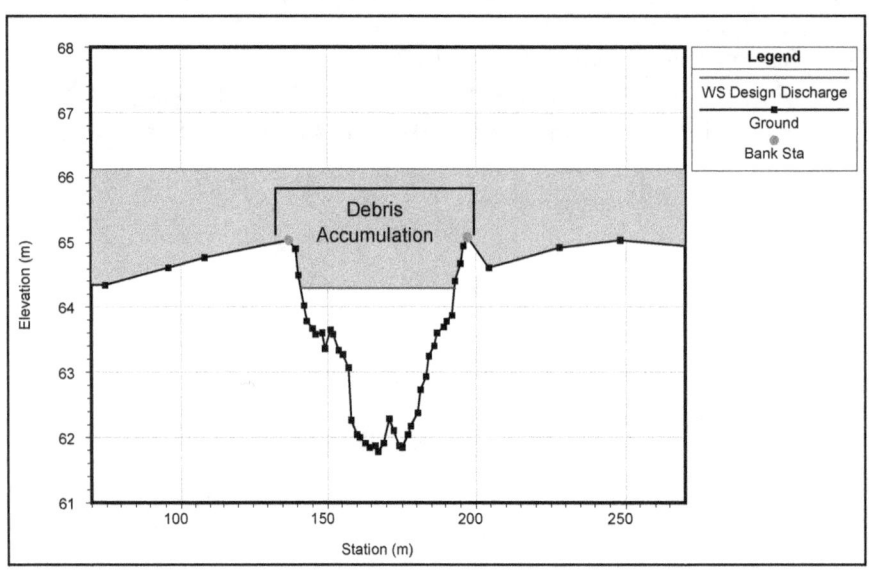

Figure 4.12. Upstream face of the bridge for Example 3.

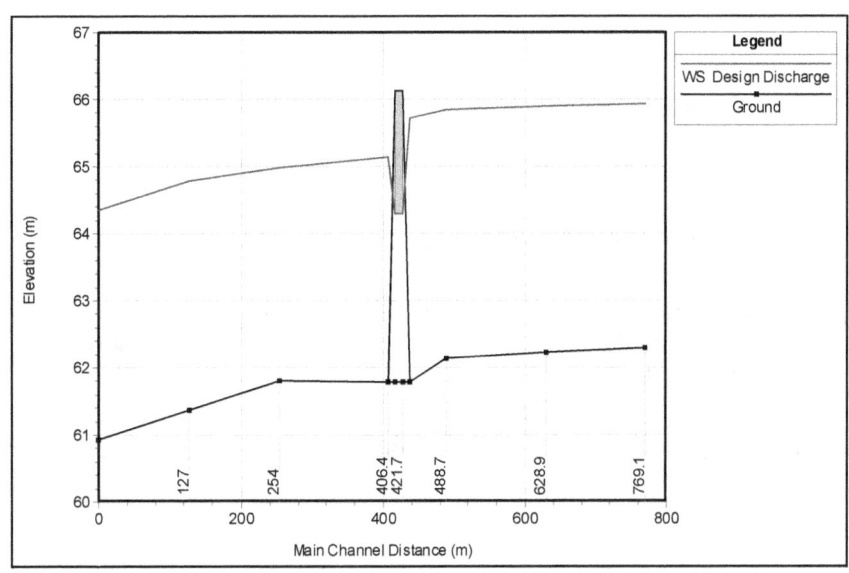

Figure 4.13. Water surface profile for Example 3.

Table 4.5. Results of Hydraulic Calculations for Example 3.

River Station (m)	Water Surface Elevation (m)	Flow Area (m^2)	Main Channel Velocity (m/s)	Cross Section Average Velocity (m/s)	Average Flow Depth[1] (m)
0	64.36	255.76	2.54	0.86	2.43
127	64.79	325.88	1.64	0.68	2.17
254	64.98	333.72	1.07	0.66	1.64
406.4	65.13	130.59	1.80	1.68	2.03
421.7 BR D	64.30	75.34	2.92	2.92	1.45
421.7 BR U	64.30	75.34	2.92	2.92	1.45
436.9	65.71	171.53	1.39	1.28	2.61
488.7	65.85	563.03	1.03	0.39	3.06
628.9	65.89	573.58	1.10	0.38	2.99
769.1	65.94	579.47	1.05	0.38	2.78

Notes:
1. For this example, the entire bridge opening was defined as the main channel. So, the average depth of the main channel is the same as the average depth of the entire cross section.

Determine:

Compute the hydrostatic and drag forces for a debris accumulation on a superstructure.

Solution:

Hydrostatic Force on Superstructure Accumulation

A_{hu} = Area of the debris accumulation below the upstream water surface
A_{hu} = (WS$_{US}$ – Debris bottom, DB$_{EL}$)(Width of debris accumulation, W$_D$)
A_{hu} = (65.71 – 64.30)(60.0) = 84.60 m^2

A_{hd} = Area of the debris accumulation below the downstream water surface
A_{hd} = (WS$_{DS}$ – DB$_{EL}$)(W$_D$)
A_{hd} = (65.13 – 64.30)(60.0) = 49.80 m^2

h_{cu} = Vertical distance to centroid of A_{hu} = 0.5(WS$_{US}$ – DB$_{EL}$)
h_{cu} = 0.5(65.71 – 64.30) = 0.71 m

h_{cd} = Vertical distance to centroid of A_{hd} = 0.5(WS$_{DS}$ – DB$_{EL}$)
h_{cd} = 0.5(65.13 – 64.30) = 0.42 m

F_{hu} = Hydrostatic force upstream = $\gamma h_{cu} A_{hu}$
F_{hu} = (9.81)(0.71)(84.60) = 589 kN

F_{hd} = Hydrostatic force downstream = $\gamma h_{cd} A_{hd}$
F_{hd} = (9.81)(0.42)(49.80) = 205 kN

F_h = Total hydrostatic force on Pile Bent 2 = F_{hu} - F_{hd}
F_h = 589 – 205 = 384 kN

Drag Force on Superstructure Accumulation

$$B = \text{Blockage ratio} = \frac{A_d}{A_d + A_c}$$

$$B = \frac{84.60}{84.60 + 75.34} = 0.53$$

B is greater than 0.3, therefore V_r should be based on the average velocity in the contracted section.

V_r = 2.92 m/s (Table 4.5)

$$Fr = \text{Froude number} = \frac{V_r}{\sqrt{g y_r}}$$

$$Fr = \frac{2.92}{\sqrt{(9.81)(1.45)}} = 0.77$$

C_D = Drag coefficient = 3.1 – 3.6B (Table 4.1)
C_D = 3.1 – (3.6)(0.53) = 1.19

$$F_D = \text{Drag force on superstructure} = C_D \gamma A_{hu} \frac{V_r^2}{2g}$$

$$F_D = (1.19)(9810)(84.60)\frac{(2.92)^2}{2(9.81)} = 429 \text{ kN}$$

Total Force on Superstructure Accumulation

F = Total segment force = $F_h + F_D$
F = 384 + 429 = 813 kN

Location of Forces on Superstructure Accumulation

$$F_{hEL} = \text{Elevation of hydrostatic force} = DB_{EL} + \frac{F_{hu}\left(\dfrac{WS_{US} - DB_{EL}}{3}\right) - F_{hd}\left(\dfrac{WS_{DS} - DB_{EL}}{3}\right)}{F_h}$$

$$F_{hEL} = 64.30 + \frac{589\left(\dfrac{65.71 - 64.30}{3}\right) - 205\left(\dfrac{65.13 - 64.30}{3}\right)}{384} = 64.87 \text{ m}$$

F_{DEL} = Elevation of drag force = $0.5(WS_{US} + DB_{EL})$
F_{DEL} = 0.5(65.71 + 64.30) = 65.00 m

$$F_{EL} = \text{Elevation of total force} = \frac{(F_D)(F_{DEL}) + (F_h)(F_{hEL})}{F}$$

$$F_{EL} = \frac{(429)(65.00) + (384)(64.87)}{813} = 64.94 \text{ m}$$

F_{hST} = Station of hydrostatic force = 0.5(Left station of debris + right station of debris)
F_{DST} = Station of drag force = F_{hST}
F_{ST} = Station of total force = F_{DST} = F_{hST}
F_{hST} = F_{DST} = F_{ST} = 0.5(137.16 + 197.21) = 167.19 m

4.7.4   Example 4 – Hydraulic Loading on a Single Pier (CU)

Given:

Design flow rate = 6,890 ft³/s
Minimum upstream main channel width = 45 ft; design log length of 45 ft
Depth of debris is full-flow depth
Main channel width at the bridge = 197 ft
Debris accumulation only on Pile Bent 2 (see Figure 4.14)
Superstructure is not submerged
Ineffective flow areas from the debris defined by 1:1 contraction and 2:1 expansion
Bottom elevation of Pile Bent 2 = 202.69 ft
Left station of debris = 526.02 ft; Right station of debris = 570.97 ft
Hydraulic computation results are provided in Table 4.6 and shown in Figure 4.15
Upstream water surface elevation, WS_{US} = 214.66 ft (Table 4.6)
Downstream water surface elevation, WS_{DS} = 213.44 ft (Table 4.6, see discussion of Downstream Water Surface Elevation in Subsection 4.3.3.1)

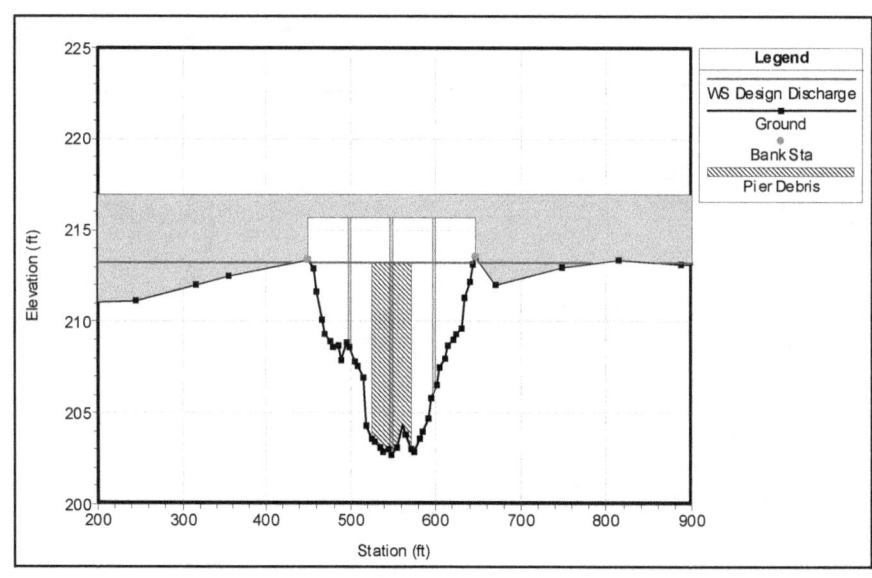

Figure 4.14. Upstream face of the bridge for Example 4.

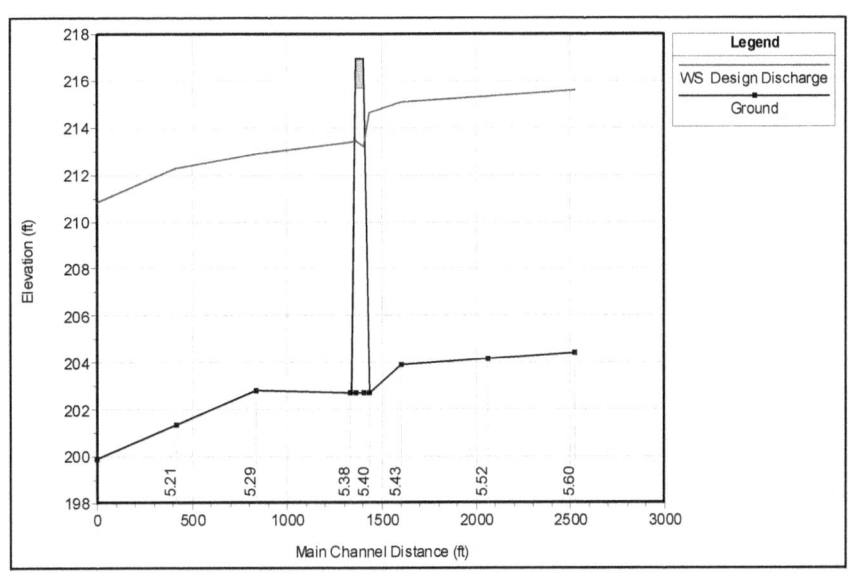

Figure 4.15. Water surface profile for Example 4.

Table 4.6. Results of Hydraulic Calculations for Example 4.

River Station (miles)	Water Surface Elevation (ft)	Flow Area (ft^2)	Main Channel Velocity (ft/s)	Cross Section Average Velocity (ft/s)	Average Flow Depth[1] (ft)
5.13	210.86	2487.36	8.12	2.77	7.69
5.21	212.27	3231.47	5.23	2.13	6.81
5.29	212.90	3312.69	3.39	2.08	5.08
5.38	213.41	1333.33	5.46	5.17	6.40
5.39 BR D	213.44	1186.32	5.81	5.81	6.36
5.39 BR U	213.21	731.66	9.42	9.42	5.21
5.40	214.66	1605.03	4.56	4.29	7.62
5.43	215.09	4472.12	4.04	1.54	9.08
5.52	215.34	4733.89	4.22	1.46	8.95
5.60	215.58	4997.62	3.91	1.38	8.36

Notes:
1. For this example, the entire bridge opening was defined as the main channel. So, the average depth of the main channel is the same as the average depth of the entire cross section.

Determine:

Compute the hydrostatic and drag forces for a debris accumulation on a single pier.

Solution:

Hydrostatic Force on Single Pier Accumulation

A_{hu} = Area of the debris accumulation below the upstream water surface
A_{hu} = (WS$_{US}$ − Debris bottom, DB$_{EL}$)(Width of debris accumulation, W$_D$)
A_{hu} = (214.66 − 202.69)(45) = 538.65 ft^2

A_{hd} = Area of the debris accumulation below the downstream water surface
A_{hd} = (WS$_{DS}$ − DB$_{EL}$)(W$_D$)
A_{hd} = (213.44 − 202.69)(45) = 483.75 ft^2

h_{cu} = Vertical distance to centroid of A_{hu} = 0.5(WS$_{US}$ − DB$_{EL}$)
h_{cu} = 0.5(214.66 − 202.69) = 5.98 ft

h_{cd} = Vertical distance to centroid of A_{hd} = 0.5(WS$_{DS}$ − DB$_{EL}$)
h_{cd} = 0.5(213.44 − 202.69) = 5.38 ft

F_{hu} = Hydrostatic force upstream = $\gamma h_{cu} A_{hu}$
F_{hu} = (62.4)(5.98)(538.65) = 200,998 lbs (100.5 tons)

F_{hd} = Hydrostatic force upstream = $\gamma h_{cd} A_{hd}$
F_{hd} = (62.4)(5.38)(483.75) = 162,401 lbs (81.2 tons)

F_h = Total hydrostatic force on Pile Bent 2 = F_{hu} - F_{hd}
F_h = 200,998 − 162,401 = 38,597 lbs (19.3 tons)

Drag Force on Single Pier Accumulation

$$B = \text{Blockage ratio} = \frac{A_d}{A_d + A_c}$$

$$B = \frac{538.65}{538.65 + 731.66} = 0.42$$

B is greater than 0.3, therefore V_r should be based on the average velocity in the contracted section.

V_r = 9.42 ft/s (Table 4.6)

$$Fr = \text{Froude number} = \frac{V_r}{\sqrt{g\,y_r}}$$

$$Fr = \frac{9.42}{\sqrt{(32.2)(5.21)}} = 0.73$$

C_D = Drag coefficient = 3.1 – 3.6B (Table 4.1)
C_D = 3.1 – (3.6)(0.42) = 1.59

$$F_D = \text{Drag force on Pile Bent 2} = C_D \gamma A_{hu} \frac{V_r^2}{2g}$$

$$F_D = (1.59)(62.4)(538.65)\frac{(9.42)^2}{2(32.2)} = 73{,}638 \text{ lbs (36.8 tons)}$$

Total Force on Single Pier Accumulation

F = Total segment force = $F_h + F_D$
F = 38,597 + 73,638 = 112,235 lbs (56.1 tons)

Location of Forces on Single Pier Accumulation

$$F_{hEL} = \text{Elevation of hydrostatic force} = DB_{EL} + \frac{F_{hu}\left(\dfrac{WS_{US} - DB_{EL}}{3}\right) - F_{hd}\left(\dfrac{WS_{DS} - DB_{EL}}{3}\right)}{F_h}$$

$$F_{hEL} = 202.69 + \frac{100.5\left(\dfrac{214.66 - 202.69}{3}\right) - 81.2\left(\dfrac{213.44 - 202.69}{3}\right)}{19.3} = 208.39 \text{ ft}$$

F_{DEL} = Elevation of drag force = $0.5(WS_{US} + DB_{EL})$
F_{DEL} = 0.5(214.66 + 202.69) = 208.68 ft

$$F_{EL} = \text{Elevation of total force} = \frac{(F_D)(F_{DEL}) + (F_h)(F_{hEL})}{F}$$

$$F_{EL} = \frac{(36.8)(208.68) + (19.3)(208.39)}{56.1} = 208.58 \text{ ft}$$

F_{hST} = Station of hydrostatic force = 0.5(Left station of debris + right station of debris)
F_{DST} = Station of drag force = F_{hST}
F_{ST} = Station of total force = $F_{DST} = F_{hST}$
$F_{hST} = F_{DST} = F_{ST}$ = 0.5(526.02 + 570.97) = 548.50 ft

4.7.5 Example 5 – Hydraulic Loading on Two Adjacent Piers, Case 1 (CU)

Given:

Design flow rate = 3,000 ft³/s

Minimum upstream main channel width = 45 ft; design log length = 45 ft

Depth of debris is full-flow depth

Main channel width at the bridge = 197 ft

Superstructure is not submerged

Ineffective flow areas from debris defined by 1:1 contraction and 2:1 expansion

Bottom elevation of Pile Bent 2 = 203.35 ft; Pile Bent 3 = 203.15 ft

Total accumulation width = 83.4 ft (defined by assuming that the accumulation extends laterally half the design log length beyond each pier).

Accumulation width on Pile Bent 2 = 45 ft for Case 1 and 38.4 ft for Case 2

Accumulation width on Pile Bent 3 = 38.4 ft for Case 1 and 45 ft for Case 2

Pile Bent 2, left station of debris = 507.53 ft; Right station of debris = 552.48 ft

Pile Bent 3, left station of debris = 552.48 ft; Right station of debris = 590.87 ft

Hydraulic computation results are provided in Table 4.7 and shown in Figure 4.17

Upstream water surface elevation, WS_{US} = 214.17 ft (Table 4.7)

Downstream water surface elevation, WS_{DS} = 211.89 ft (Table 4.7, see discussion of <u>Downstream Water Surface Elevation</u> in Subsection 4.3.3.1)

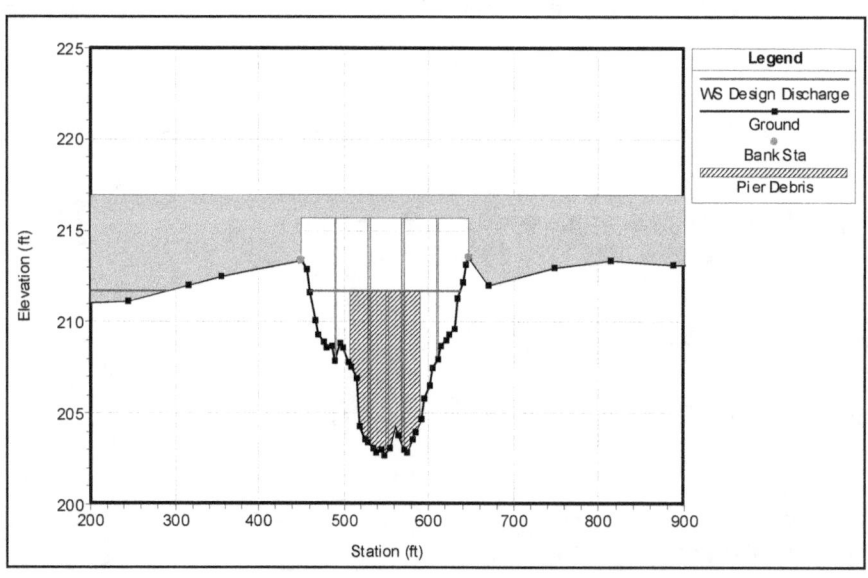

Figure 4.16. Upstream face of the bridge for Example 5.

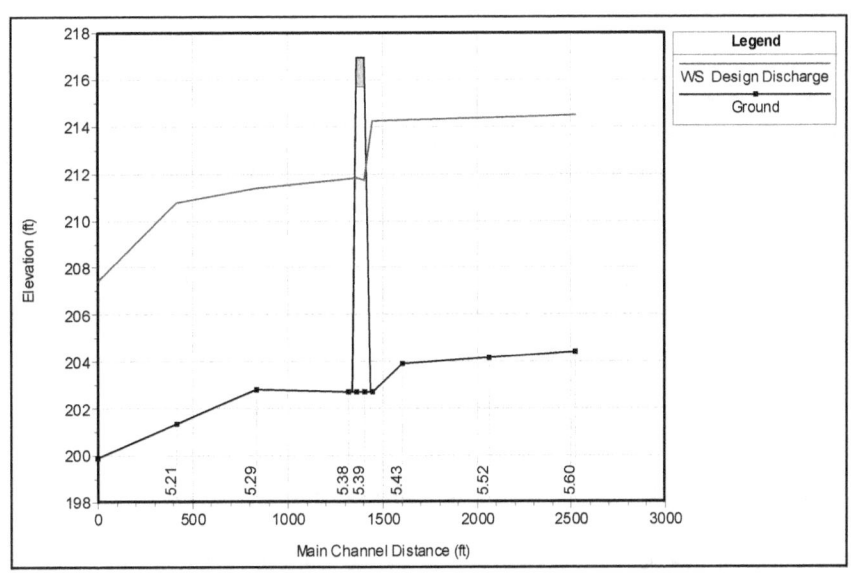

Figure 4.17. Water surface profile for Example 5.

Table 4.7. Results of Hydraulic Calculations for Example 5.

River Station (miles)	Water Surface Elevation (ft)	Flow Area (ft^2)	Main Channel Velocity (ft/s)	Cross Section Average Velocity (ft/s)	Average Flow Depth[1] (ft)
5.13	207.42	310.14	11.55	9.68	4.53
5.21	210.80	1894.63	4.00	1.57	5.35
5.29	211.36	1897.97	2.69	1.57	3.54
5.38	211.79	974.12	3.15	3.08	5.35
5.39 BR D	211.82	898.55	3.35	3.35	5.35
5.39 BR U	211.76	295.07	10.17	10.17	3.25
5.40	214.25	1530.45	2.10	1.97	7.22
5.43	214.31	3325.62	2.26	0.92	8.30
5.52	214.41	3322.07	2.59	0.92	8.01
5.60	214.51	3392.69	2.56	0.89	7.28

Notes:
1. For this example, the entire bridge opening was defined as the main channel. So, the average depth of the main channel is the same as the average depth of the entire cross section.

4.33

Determine:

Compute the hydrostatic and drag forces for a debris accumulation on two piers. The calculations are for only Case 1, which is based on the width of the debris accumulation on Pile Bent 2 being equal to the design log length and Pile Bent 3 having a smaller accumulation width. Case 2 is the reverse of Case 1, i.e., the width of the accumulation on Pile Bent 3 would be equal to the design log length and Pile Bent 2 would have a smaller accumulation width. Case 3 is based on the assumption that the design log length spans the opening approximately in the middle of the two piers and the resulting load on the accumulation is transferred equally to each pier.

Solution:

Hydrostatic Forces on Two Adjacent Piers (Case 1)

A_{hu} = Area of the debris accumulation below the upstream water surface
A_{hu} = (WS_{US} – Debris bottom, DB_{EL})(Width of debris accumulation, W_D)
A_{hu} = (214.25 – 203.35)(45) = 490.50 ft^2 for Pile Bent 2
A_{hu} = (214.25 – 203.15)(38.4) = 426.24 ft^2 for Pile Bent 3

A_{hd} = Area of the debris accumulation below the downstream water surface
A_{hd} = (WS_{DS} – DB_{EL})(W_D)
A_{hd} = (211.82 – 203.35)(45) = 381.15 ft^2 for Pile Bent 2
A_{hd} = (211.82 – 203.15)(38.4) = 332.93 ft^2 for Pile Bent 3

h_{cu} = Vertical distance to centroid of A_{hu} = 0.5(WS_{US} – DB_{EL})
h_{cu} = 0.5(214.25 – 203.35) = 5.45 ft   for Pile Bent 2
h_{cu} = 0.5(214.25 – 203.15) = 5.55 ft   for Pile Bent 3

h_{cd} = Vertical distance to centroid of A_{hd} = 0.5(WS_{DS} – DB_{EL})
h_{cd} = 0.5(211.82 – 203.35) = 4.24 ft   for Pile Bent 2
h_{cd} = 0.5(211.82 – 203.15) = 4.34 ft   for Pile Bent 3

F_{hu} = Hydrostatic force upstream = $\gamma h_{cu} A_{hu}$
F_{hu} = (62.4)(5.45)(490.50) = 166,809 lbs (83.4 tons)   for Pile Bent 2
F_{hu} = (62.4)(5.55)(426.24) = 147,615 lbs (73.8 tons)   for Pile Bent 3

F_{hd} = Hydrostatic force upstream = $\gamma h_{cd} A_{hd}$
F_{hd} = (62.4)(4.24)(381.15) = 100,843 lbs (50.4 tons)   for Pile Bent 2
F_{hd} = (62.4)(4.34)(332.93) = 90,163 lbs (45.1 tons)   for Pile Bent 3

F_h = Total hydrostatic force on Pile Bent 2 = F_{hu} - F_{hd}
F_h = 166,809 – 100,843 = 65,966 lbs (33.0 tons)   for Pile Bent 2
F_h = 147,615 – 90,163 = 57,452 lbs (28.7 ton)   for Pile Bent 3

Drag Force on Two Adjacent Piers (Case 1)

$$B = \text{Blockage Ratio} = \frac{A_d}{A_d + A_c} = \frac{A_{hu\,pier\,2} + A_{hu\,pier\,3}}{A_{hu\,pier\,2} + A_{hu\,pier\,3} + A_c}$$

$$B = \frac{490.50 + 426.24}{490.50 + 426.24 + 295.07} = 0.76$$

B is greater than 0.3, therefore V_r should be based on the average velocity in the contracted section.

V_r = 10.17 ft/s (Table 4.7)

$$Fr = \text{Froude number} = \frac{V_r}{\sqrt{g\,y_r}}$$

$$Fr = \frac{10.17}{\sqrt{(32.2)(3.25)}} = 0.99$$

C_D = Drag coefficient = 3.1 – 3.6B (Table 4.1)
C_D = 3.1 – (3.6)(0.76) = 0.36

$$F_D = \text{Drag force on Pile Bent} = C_D \gamma A_{hu} \frac{V_r^2}{2g}$$

$$F_D = (0.36)(62.4)(490.50)\frac{(10.17)^2}{2(32.2)} = 19{,}589 \text{ lbs (9.8 tons)} \quad \text{for Pile Bent 2}$$

$$F_D = (0.36)(62.4)(426.24)\frac{(10.17)^2}{2(32.2)} = 17{,}023 \text{ lbs (8.5 tons)} \quad \text{for Pile Bent 3}$$

Total Force on Two Adjacent Piers (Case 1)

F = Total segment force = $F_h + F_D$
F = 65,966 + 19,589 = 85,555 lbs (42.8 tons)   for Pile Bent 2
F = 57,452 + 17,023 = 74,475 lbs (37.2 tons)   for Pile Bent 3

Location of Forces on Single Pier Accumulation

$$F_{hEL} = \text{Elevation of hydrostatic force} = DB_{EL} + \frac{F_{hu}\left(\dfrac{WS_{US} - DB_{EL}}{3}\right) - F_{hd}\left(\dfrac{WS_{DS} - DB_{EL}}{3}\right)}{F_h}$$

$$F_{hEL} = 203.35 + \frac{83.40\left(\dfrac{214.25 - 203.35}{3}\right) - 50.4\left(\dfrac{211.82 - 203.35}{3}\right)}{33.0} = 208.22 \text{ ft} \quad \text{for Pile Bent 2}$$

$$F_{hEL} = 203.15 + \dfrac{73.8\left(\dfrac{214.25 - 203.15}{3}\right) - 45.10\left(\dfrac{211.82 - 203.15}{3}\right)}{28.7} = 208.12 \text{ ft} \quad \text{for Pile Bent 3}$$

F_{DEL} = Elevation of drag force = $0.5(WSEL_{US} + DBEL)$
$F_{DEL} = 0.5(214.25 + 203.35) = 208.80$ ft   for Pile Bent 2
$F_{DEL} = 0.5(214.25 + 203.15) = 208.70$ ft   for Pile Bent 3

$$F_{EL} = \text{Elevation of total force} = \frac{(F_D)(F_{DEL}) + (F_h)(F_{hEL})}{F}$$

$$F_{EL} = \frac{(9.8)(208.80) + (33.0)(208.22)}{42.8} = 208.35 \text{ ft} \quad \text{for Pile Bent 2}$$

$$F_{EL} = \frac{(8.5)(208.70) + (28.7)(208.12)}{37.2} = 208.25 \text{ ft} \quad \text{for Pile Bent 3}$$

F_{hST} = Station of hydrostatic force = 0.5(Left station of debris + right station of debris)
F_{DST} = Station of drag force = F_{hST}
F_{ST} = Station of total force = $F_{DST} = F_{hST}$
$F_{hST} = F_{DST} = F_{ST} = 0.5(507.53 + 552.48) = 530.00$ ft   for Pile Bent 2
$F_{hST} = F_{DST} = F_{ST} = 0.5(552.48 + 590.87) = 571.67$ ft   for Pile Bent 3

4.7.6   Example 6 – Hydraulic Loading on a Superstructure (CU)

Given:

Design flow rate = 7,770 ft³/s
Low Chord Elevation = 214.91 ft
Depth of debris is 3.94 feet below the bridge low chord = 210.97 ft
Debris accumulation extends along the entire length of the structure (see Figure 4.18)
Main channel width at the bridge = 197 ft
Left station of debris = 450.00 ft; Right station of debris = 647.01 ft
Hydraulic computation results are provided in Table 4.8 and shown in Figure 4.19
Upstream water surface elevation, WS_{US} = 215.59 ft (Table 4.8)
Downstream water surface elevation, WS_{DS} = 213.69 ft (Table 4.8, see discussion of Downstream Water Surface Elevation in Subsection 4.3.3.1)

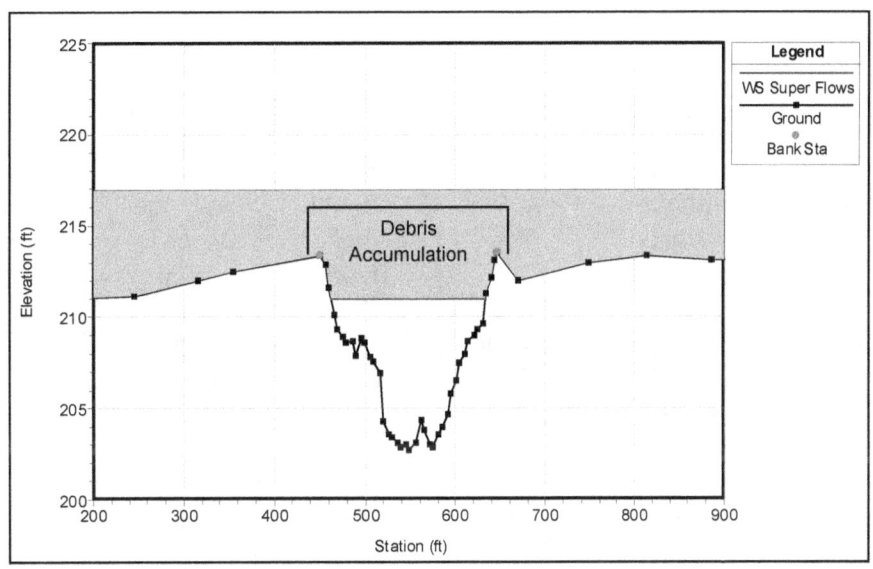

Figure 4.18. Upstream face of the bridge for Example 6.

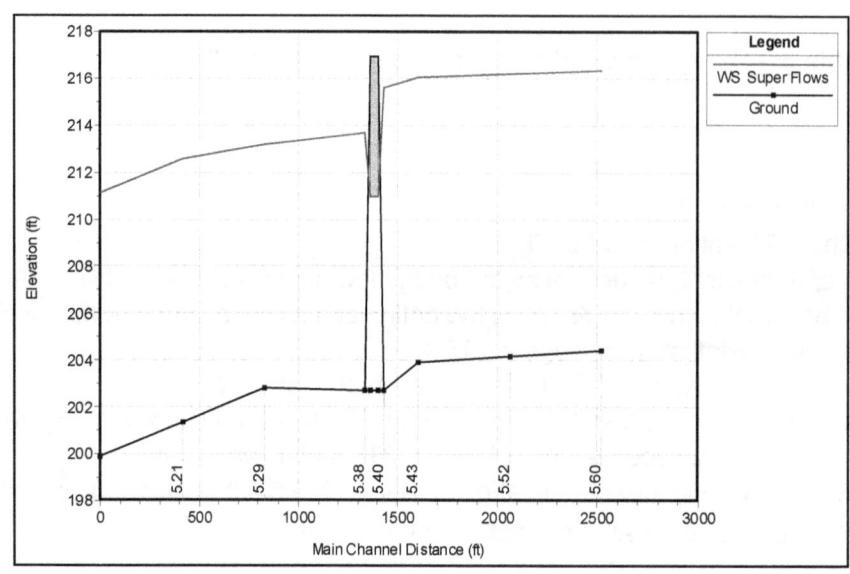

Figure 4.19. Water surface profile for Example 6.

Table 4.8. Results of Hydraulic Calculations for Example 6.

River Station (miles)	Water Surface Elevation (ft)	Flow Area (ft^2)	Main Channel Velocity (ft/s)	Cross Section Average Velocity (ft/s)	Average Flow Depth Depth[1] (ft)
5.13	211.17	2753.25	8.33	2.82	7.97
5.21	212.58	3508.09	5.38	2.23	7.12
5.29	213.20	3592.48	3.51	2.17	5.38
5.38	213.69	1405.80	5.91	5.51	6.66
5.39 BR D	210.97	811.03	9.58	9.58	4.76
5.39 BR U	210.97	811.03	9.58	9.58	4.76
5.40	215.59	1846.51	4.56	4.20	8.56
5.43	216.05	6061.00	3.38	1.28	10.04
5.52	216.19	6174.57	3.61	1.25	9.81
5.60	216.35	6237.97	3.45	1.25	9.12

Notes:
1. For this example, the entire bridge opening was defined as the main channel. So, the average depth of the main channel is the same as the average depth of the entire cross section.

Determine:

Compute the hydrostatic and drag forces for a debris accumulation on a superstructure.

Solution:

Hydrostatic Force on Superstructure Accumulation

A_{hu} = Area of the debris accumulation below the upstream water surface
A_{hu} = (WS_{US} – Debris bottom, DB_{EL})(Width of debris accumulation, W_D)
A_{hu} = (215.59 – 210.97)(197.0) = 910.14 ft^2

A_{hd} = Area of the debris accumulation below the downstream water surface
A_{hd} = (WS_{DS} – DB_{EL})(W_D)
A_{hd} = (213.69 – 210.97)(197.0) = 535.84 ft^2

h_{cu} = Vertical distance to centroid of A_{hu} = 0.5(WS_{US} – DB_{EL})
h_{cu} = 0.5(215.59 – 210.97) = 2.31 ft

h_{cd} = Vertical distance to centroid of A_{hd} = 0.5(WS_{DS} – DB_{EL})
h_{cd} = 0.5(213.69 – 210.97) = 1.36 ft

F_{hu} = Hydrostatic force upstream = $\gamma h_{cu} A_{hu}$
F_{hu} = (62.4)(2.31)(910.14) = 131,191 lbs (65.6 tons)

F_{hd} = Hydrostatic force upstream = $\gamma h_{cd} A_{hd}$
F_{hd} = (62.4)(1.36)(535.84) = 45,474 lbs (22.7 tons)

F_h = Total hydrostatic force on Pile Bent 2 = F_{hu} - F_{hd}
F_h = 131,191 – 45,474 = 85,717 lbs (42.9 tons)

Drag Force on Superstructure Accumulation

$$B = \text{Blockage ratio} = \frac{A_d}{A_d + A_c}$$

$$B = \frac{910.40}{910.40 + 811.03} = 0.53$$

B is greater than 0.3, therefore V_r should be based on the average velocity in the contracted section.

V_r = 9.58 ft/s (Table 4.8)

$$Fr = \text{Froude number} = \frac{V_r}{\sqrt{gy_r}}$$

$$Fr = \frac{9.58}{\sqrt{(32.2)(4.76)}} = 0.77$$

C_D = Drag coefficient = 3.1 – 3.6B (Table 4.1)
C_D = 3.1 – (3.6)(0.53) = 1.19

$$F_D = \text{Drag force on superstructure} = C_D \gamma A_{hu} \frac{V_r^2}{2g}$$

$$F_D = (1.19)(62.4)(910.14)\frac{(9.58)^2}{2(32.2)} = 96{,}313 \text{ lbs (48.1 tons)}$$

Total Force on Superstructure Accumulation

F = Total segment force = $F_h + F_D$
F = 85,717 + 96,313 = 182,030 lbs (91.0 tons)

Location of Forces on Superstructure Accumulation

$$F_{hEL} = \text{Elevation of hydrostatic force} = DB_{EL} + \frac{F_{hu}\left(\frac{WS_{US} - DB_{EL}}{3}\right) - F_{hd}\left(\frac{WS_{DS} - DB_{EL}}{3}\right)}{F_h}$$

$$F_{hEL} = 210.97 + \frac{65.6\left(\frac{215.59 - 210.97}{3}\right) - 22.07\left(\frac{213.69 - 210.97}{3}\right)}{42.9} = 212.85 \text{ ft}$$

F_{DEL} = Elevation of drag force = $0.5(WS_{US} + DB_{EL})$
F_{DEL} = 0.5(215.59 + 210.97) = 213.28 ft

$$F_{EL} = \text{Elevation of total force} = \frac{(F_D)(F_{DEL}) + (F_h)(F_{hEL})}{F}$$

$$F_{EL} = \frac{(48.1)(213.28) + (42.9)(212.85)}{91.0} = 213.08 \text{ ft}$$

F_{hST} = Station of hydrostatic force = 0.5(Left station of debris + right station of debris)
F_{DST} = Station of drag force = F_{hST}
F_{ST} = Station of total force = F_{DST} = F_{hST}
$F_{hST} = F_{DST} = F_{ST}$ = 0.5(450.00 + 647.01) = 548.51 ft

CHAPTER 5 – DEBRIS COUNTERMEASURES

5.1   BACKGROUND

Countermeasures to mitigate and protect effects of debris depend on the type of structure. Typically, these countermeasures are grouped into structural and non-structural measures. The structural measures have many configurations and constructed from many materials. The non-structural measures typically involve long-term approaches.

This Chapter will describe debris countermeasures for culverts and bridges. As will be described in this and subsequent Chapters, while some countermeasures may have some applicability to both types of structures (as well as other not described herein), engineering judgment on use remains a key design consideration.

5.2   COUNTERMEASURES FOR CULVERTS

5.2.1   <u>Structural Measures</u>

There are various types of structural measures available for culverts. These measures can have many shapes and can be constructed using various materials. The measures can generally be divided into the following types:

Debris Deflectors are structures placed at the culvert inlet to deflect the major portion of the debris away from the culvert entrance. They are normally "V"-shaped in plan with the apex upstream. Examples of this type of structure measure are shown in Figure 5.1 through Figure 5.9.

Debris Racks are structures placed across the stream channel to collect the debris before it reaches the culvert entrance. Debris racks are usually vertical and at right angles to the streamflow, but they may be skewed with the flow or inclined with the vertical. Pictures of debris racks are shown in Figure 5.10 through Figure 5.22.

Debris Risers are a closed-type structure placed directly over the culvert inlet to cause deposition of flowing debris and fine detritus before it reaches the culvert inlet. Risers are usually built of metal pipe. Examples of debris risers are shown in Figure 5.23 through Figure 5.25. Risers can also be used as relief devices in the event the entrance becomes completely blocked with debris (Figure 5.25).

Debris Cribs are open crib-type structures placed vertically over the culvert inlet in log-cabin fashion to prevent inflow of coarse bed load and light floating debris. Photos of this type of structure are provided in Figure 5.26 through Figure 5.28.

Debris Fins are walls built in the stream channel upstream of the culvert. Their purpose is to align the debris with the culvert so that the debris would pass through the culvert without

5.1

accumulating at the inlet. This type of measure can also be used at bridge. Examples of this type of structure measure for culverts are shown in Figure 5.29 through Figure 5.35.

Debris Dams and Basins are structures placed across well-defined channels to form basins which impede the stream flow and provide storage space for deposits of detritus and floating debris. This type of structure is shown in Figure 5.36 through Figure 5.39.

Combination Devices are a combination of two or more of the preceding debris-control structures at one site to handle more than one type of debris and to provide additional insurance against the culvert inlet from becoming clogged. Examples of combination devices are shown in Figure 5.40.

5.2.2 Non-structural Measures

The only type of non-structural measures available for culvert structures is to provide emergency and annual maintenance. Although not always feasible for remote culverts or culverts with small drainage areas, maintenance could be a viable option for larger culverts with fairly large drainage basins. Emergency maintenance could involve removing debris from the culvert entrance and/or an existing debris-control structure. Annual maintenance could involve removing debris from within the culvert, at the culvert entrance, and/or immediately upstream of the culvert, or repairing any existing structural measures.

5.3 COUNTERMEASURES FOR BRIDGES

5.3.1 Structural Measures

Various types of structural measures are also available for bridge structures. Some of the measures discussed above for the culvert structures can also be utilized at bridges. The various types include:

Debris Fins are walls built in the stream channel upstream of the bridge to align large floating trees so that their length is parallel to the flow, enabling them to pass under the bridge without incident. This type of measure is also referred to as a "pier nose extension". Examples of debris fin deflectors are provided in Figure 5.35.

In-channel Debris Basins are structures placed across well-defined channels to form basins which impede the streamflow and provide storage space for deposits of detritus and floating debris. These structures can be expensive to construct and maintain. This type of structure is shown in Figure 5.36 through Figure 5.40.

River-Training Structures are structures placed in the river flow to create counter-rotating streamwise vortices in their wakes to modify the near-bed flow pattern to redistribute flow and sediment transport within the channel cross section. Examples of this type of structure include

Iowa vanes, and impermeable and permeable spurs. This type of structure is shown in Figure 5.41 and Figure 5.42.

Crib Structures are walls built between open-pile bents to prevent debris lodging between the bents. The walls are typically constructed out of timber or metal material.

Flood Relief Sections are overtopping or flow through structures that divert excess flow and floating debris away from the bridge structure and through the structure.

Debris Deflectors are structures placed upstream of the bridge piers to deflect and guide debris through the bridge opening. They are normally "V"-shaped in plan with the apex upstream. An example of this type of structure is shown in Figure 5.43. A special type of debris deflector is a hydrofoil. Hydrofoils are submerged structures placed immediately upstream of bridge piers that create counter-rotating streamwise vortices in their wakes to deflect and divert floating debris around the piers and through the bridge opening. Unfortunately, no hydrofoils have been implemented within the field. They have only been tested within a physical model study.[56]

Debris Sweeper is a polyethylene device that is attached to a vertical stainless steel cable or column affixed to the upstream side of the bridge pier. The polyethylene device travels vertically along the pier as the water surface rises and falls. It is also rotated by the flow, causing the debris to be deflected away from the pier and through the bridge opening. This type of device is shown in Figure 5.44 through Figure 5.47.

Booms are logs or timbers that float on the water surface to collect floating drift. Drift booms require guides or stays to hold them in place laterally. Booms are very limited in use and their application is not covered within this manual.

Design Features are structural features that can be implemented in the design of a proposed bridge structure. The first feature is freeboard, which is a safety precaution of providing additional space between the maximum water surface elevation and the low chord elevation of the bridge. The second feature is related to the type of piers and the location and spacing of the piers. Ideally, the piers should be a solid wall type pier that is aligned with the approaching flow. They should also be located and spaced such that the potential for debris accumulation is minimized. The third feature involves the use of special superstructure design, such as thin decks, to prevent or reduce the debris accumulation on the structure when the flood stage rises above the deck. The last feature involves providing adequate access to the structure for emergency and annual maintenance.

5.3.2 Non-Structural Measures

There are generally two types of non-structural measures available for bridge structures. The first type of non-structural measure is emergency and annual maintenance. Emergency maintenance could involve removing debris from the bridge piers and/or abutments; placing riprap near the piers, abutments, or where erosion is occurring due to flow impingement created by the debris accumulation; and/or dredging of the channel bottom. Annual maintenance could involve debris removal and repair to any existing structural measures.

The second type of non-structural measure is management of the upstream watershed. The purpose of this measure is to reduce the amount of debris delivered to the structure by reducing the sources of debris, preventing the debris from being introduced into the streams, and clearing debris from the stream channels. The type of management system implemented varies depending on the type of debris. For organic floating debris, the management system could involve removing dead and decayed trees, and/or debris jams; providing buffer zones for areas where logging practices exist; implementing a cable-assisted felling of trees system; and stabilizing hillside slopes and stream banks.

5.4   COUNTERMEASURES FOR FIRE DAMAGED / DEFORESTED AREAS

5.4.1   Fire Damaged Areas

Fires can decrease the amount of floating debris introduced into the stream system. However, fires increase the magnitude of runoff from the burned area, increase the erodibility of soils, and increase the probability of catastrophic events such as debris flows and landslides, resulting in a significant increase in sediment yield from the effected area. This increase could cause an increase in fine and coarse detritus to be transported to and deposited at a culvert or bridge structure. Countermeasures that can be implemented to reduce the amount of material transported to a drainage structure include:

Surface Treatments are countermeasures that are placed directly on the burnt landscape to reduce the potential for erosion from the disturbed area. There are various types of surface treatments. One type of surface treatment is hydroseeding, which involves re-vegetation of the landscape by spraying grass or wildflower seeds. This method can be easily applied to large areas, and it is most effective when there is adequate time for the vegetation to develop. Another type of surface treatment consists of placing straw or wood fiber mulch on the landscape. A fabric mat can be used in lieu of mulch material to provide more resistance to erosive forces.

Sediment barriers are temporary structures used to help retain the soil on the site and reduce the runoff velocity across areas below it. One type of sediment barrier is a silt fence, which is a temporary structures of wood or steel fend posts, weir mesh fencing, and a suitable permeable filter fabric. Another type of a sediment barrier structure is a straw bale dike, which are constructed out of straw bales. Both of these structures should be limited to small drainage areas that have a maximum slope of 2H on 1V and flow path length of around 100 feet. Another type of sediment barrier is straw wattles. Wattles are tubes of straw or coconut fiber. Wattles help stabilize the slope by shortening the slope length and by slowing, spreading and filtering overland water flow. They are placed in trenches on the slope at selected vertical spacing and held in-place by stakes.

In-channel Debris Basins are structures placed across well-defined channels to form basins which impede the streamflow and provide storage space for deposits of detritus and floating debris.

5.4.2 Deforestation

Logging practices can cause a substantial increase in the volume of floating debris entering a channel system. Practices that reduce the quantities of floating debris include directional felling uphill with a tree-pulling system and providing a buffer strip of undisturbed vegetation along the streams. As in fires, logging can cause an increase in magnitude of runoff from the disturbed area, increase the erodibility of soils, and increase the probability of catastrophic events such as debris flows and landslides, resulting in an increase in fine and coarse detritus to be transported to and deposited at a drainage structure. The countermeasures that can be implemented to reduce the amount of material transported to a drainage structure include sediment barriers and in-channel debris basins as discussed above for fires.

Figure 5.1. Steel rail debris deflector for large rock (looking upstream of culvert).

Figure 5.2. Steel rail debris deflector (looking downstream).

Figure 5.3. Steel rail and cable debris deflector. In boulder areas, cable is more desirable for its flexibility than a rigid rail (looking towards entrance).

Figure 5.4. Steel debris deflectors installed at entrances to a battery of culverts.

Figure 5.5. Steel rail debris deflector for battery of culverts (see Figure 5.6).

Figure 5.6. Installation of Figure 5.5 during flood; functions well under heavy debris flow.

Figure 5.7. Steel rail debris deflector in area of heavy flowing debris (looking upstream).

Figure 5.8. Timber pile debris deflector for boulders and large floating debris.

Figure 5.9. Timber pile debris deflector protected culvert during heavy floods. Nearby culverts without deflectors were plugged.

Figure 5.10. Rail debris rack over sloping inlet. Heavy debris and boulders ride over rack and leave flow to culvert unimpeded.

Figure 5.11. Post and rail debris rack, in place for 35 years, for light to medium floating debris installed 100 ft upstream of culvert.

Figure 5.12. Rail debris rack.

Figure 5.13. Timber debris rack (note how suspended by cables).

Figure 5.14. Hinged steel debris rack in urban area. Due to nature of debris and possible entry by children, bar spacing is close.

Figure 5.15. Steel debris rack in urban area.

Figure 5.16. Debris rack used in State of Washington.

Figure 5.17. Rail debris rack in arid region (see Figure 5.18).

Figure 5.18. Installation in Figure 5.17 after several years of fine silt deposition at entrance.

Figure 5.19. Steel rail debris rack. Note amount of debris accumulation in upstream channel.

Figure 5.20. Steel debris rack probably saved the culvert from plugging.

Figure 5.21. Steel grill debris rack with provision for cleanout afforded by concrete paved area in foreground.

Figure 5.22. Steel grill debris rack on slope mitered culvert entrance.

Figure 5.23. Metal pipe debris riser in basin (note anti-vortex device on top).

Figure 5.24. Metal pipe debris riser placed during initial construction of culvert provides relief in case the culvert entrance becomes plugged (see Figure 5.25).

Figure 5.25. Installation shown in Figure 5.24 after flood. Riser conveyed large flows during flood. Fence partially surrounding riser was of no value for debris control.

Figure 5.26. Debris crib of precast concrete sections and metal dowels. Height increased by extending dowels and adding more sections.

Figure 5.27. Arid region debris crib of precast concrete sections and metal dowels.

Figure 5.28. Redwood debris crib with spacing to prevent passage of fine material. Basin had buildup of 30 feet.

Figure 5.29. Concrete debris fins with sloping leading edge as extension of culvert walls.

Figure 5.30. Concrete debris fin with sloping leading edge as extension of center wall.

Figure 5.31. Concrete debris fin with rounded vertical leading edge as extension of culvert center wall.

Figure 5.32. Combined installation of concrete debris fin and metal pipe debris riser with single corrugated metal pipe culvert (looking downstream).

Figure 5.33. Concrete debris fin for single culvert (Prefer more area between wingwalls and fin).

Figure 5.34. Debris fin and metal pipe debris riser in conjunction with single barrel culvert.

Figure 5.35. Timber debris fins with sloping leading edge.

Figure 5.36. Metal bin type debris dam.

Figure 5.37. Gabion debris dam.

Figure 5.38. Debris dam of precast concrete sections fabricated to enable placement in interlocking fashion.

Figure 5.39. Debris dam of precast concrete sections fabricated to enable placement in interlocking fashion.

Figure 5.40. Debris dam and basin along with steel debris rack over culvert entrance in foreground. A metal pipe riser is visible over the spillway.

Figure 5.41. Bendway wiers on outer bank of Hatchie River looking upstream (TDOT).

Figure 5.42. Kellner jacks used for redirecting the flow patterns.

Figure 5.43. Debris deflectors installed at State Route 59 south crossing of the Eel River in central Indiana.

Figure 5.44. Debris sweeper being installed on a bridge over Staunton River in Altavista, Virginia.

Figure 5.45. Close up of a debris sweeper installed on the Cedar Creek in Washington.

Figure 5.46. Close up of a debris sweeper installed on the South Fork Obion River in Tennessee.

Figure 5.47. Close up of double-stacked installation debris sweeper on Interstate 24 over the Mississippi River.

CHAPTER 6 – DESIGN PROCEDURES FOR DEBRIS COUNTERMEASURES

6.1   GENERAL PROCEDURES AND CONSIDERATIONS

6.1.1   Field Investigations

Field investigations should be conducted prior to the design of a debris-control countermeasure or culvert/bridge structure. The purpose of the investigations is obtaining a general understanding of the debris problem at the site; acquiring data required for estimating the quantities of debris transported to the site; performing hydrologic, hydraulic, and sedimentation analyses; and attaining other miscellaneous design data. Several field investigations may be required for obtaining all of the data.

The type of debris transported to the site will influence the selection of the debris-control countermeasure, and define the type of data and analyses required to estimate the quantity of debris transported to the site. The estimated quantity of debris is needed by the designer to provide adequate debris storage immediately upstream of the site or to evaluate the potential impacts associated with debris accumulating on the culvert/bridge structure. The most useful and desired source of information on the types and quantities of debris delivered to the site would be from past floods. Such information could be secured from maintenance personnel, from inhabitants of the immediate area, or by personal observation. Unfortunately, this type of information rarely exists. Therefore, information to assist the designer in estimating the types and quantities of debris transported to the site must be obtained from field investigations. This information may include soil, land use, and topographic mapping; additional survey data; stream and watershed characteristics upstream of the site; aerial photographs; observations of the flow characteristics near the site and any direct and indirect evidence of high delivery potential for floating debris upstream of the site; sediment and discharge data; and future changes in the watershed.

Land use and soil maps are useful in estimating sediment yields of fine and coarse sediment. Land use maps can also provide an indication of future changes in the watershed that might influence the quantities of debris delivered to the site. There are many uses for topographic mapping and survey data (surveyed cross sections or digital terrain models, DTM). Some of these uses include developing a hydraulic model to evaluate the hydraulic characteristics upstream and downstream of the structure, defining the flow path of floating debris, defining the maximum allowable headwater elevation for a culvert structure, estimating the amount of debris storage available at the site, and defining potential access to the site. Sediment yield rates from gully, channel bank erosion, and mass wasting can be estimated by making a comparison between existing and historical topographic maps or survey data.

Information on the stream and watershed characteristics would include the locations and approximate extent of any lateral channel instabilities, aggradation or degradation trends within the watershed, roughness coefficients of the main channel and floodplain, type of sediment in the streambed and banks, location of any hydraulic controls, high water marks, channel dimensions, locations of existing debris accumulations, vegetation characteristics, and any potential damage locations if the debris-control countermeasure and/or culvert/bridge structure becomes clogged. Most of this information could be obtained when applying the stream reconnaissance technique documented in Chapter 4 of HEC-20 [34].

A well documented methodology for predicting bank erosion associated with stream meander migration using aerial photograph and maps has been developed by Lagasse, Spitz, Zevenbergen, Zachmann, and Thorne[36,37]. The research for the development of this methodology was conducted as part of the National Cooperative Highway Research Program (NCHRP Project 24-16) for the Transportation Research Board, and the principal product of the research is a stand-alone Handbook[37]. Information covered by the Handbook[37] includes:

- **Screening and classification of meander sites.** A morphological classification scheme for alluvial rivers developed by Brice[11] was selected as the most appropriate system for the methodology developed. The original classification, however, was modified since not all of the original classes are commonly encountered. Based on the modified classification system, a screening system was established to define if the methodology would be applicable to a study reach. The methodology is not applicable for the classifications that possess either considerable stability or excessive instability.

- **Sources of mapping and aerial photographic data.** The Handbook provides guidance on obtaining contemporary and historical aerial photographs and maps.

- **Basic principles and theory of aerial photograph comparison.** The Handbook includes a brief discussion on the types of photogrammetry, photogrammetric products, and the application of photogrammetry to meander migration.

- **Manual overlay and computer assisted techniques.** The procedures for the manual overlay techniques documented in the Handbook are briefly presented as follows: (1) obtain aerial photograph and maps; (2) convert aerials/maps to a common scale; (3) define common points for all aerial/maps; (4) trace banklines and registration points onto a transparent overlay for each set of aerials/maps; (5) define the average bankline arc, the radius of curvature of the bend, and the bend centroid position for best-fitted circles of the outer bank of each bend; and (6) define the future position of the bend by simple extrapolation based on the assumption that the bend will continue to move at the same rate and in approximately the same direction as it has in the past. The same general procedure for the manual overlay techniques can be accomplished more easily and efficiently using common computer software with drawing capabilities, such as most word processing and presentation applications or more powerful and versatile computer aided drawing (CAD) programs.

- **GIS-based measurement and extrapolation techniques.** Another product of the research was the development of Geographic Information System (GIS) software based menu-driven extensions to assist in applying the developed methodology. Two extensions were developed: (1) Data Logger, and (2) Channel Migration Predictor. The Data Logger extension was developed to streamline the measurement and analysis of bend migration data and aid in predicting channel migration. This extension provides the user a quick and easy way to gather and archive river planform data. This extension records various river characteristics that are arranged by the river reach for each river bend and historical record. The data archived by this extension is then utilized by the Channel Migration Predictor extension to predict the probable magnitude and direction of the bend migration at some specified time in the future. Both of these extensions are provided on a CD in Appendix G of the Handbook.

- **Statistical analysis.** A statistical analysis of an extensive data set (nearly 2,500 measurements) was conducted to determine relationship for channel migration to the morphological classification scheme for alluvial rivers. This analysis showed that equiwidth meander reaches are fairly stable and will not change significantly over time, applying standard regression techniques to directly predict meander migration did not yield statistically significant relationships, and the results from the frequency approach should be used primarily as a supplement to the comparative analysis using aerial photography.

- **Sources of error and limitations.** Information related to the sources of errors and limitations for using the developed methodology is provided in Chapter 6 of the Handbook.

- **Illustrated examples and application using manual overlay techniques.** An illustrated example for applying the developed methodology is provided in Chapter 8 of the Handbook.

Aerial photographs are useful in defining the roughness coefficients of the stream and floodplain, defining the representative reaches for estimating the volume of floating debris within a watershed as discussed in Chapter 3 of this manual, defining locations of bank erosion, and showing the present conditions of the watershed. Sequences of historical aerial photographs can be used to estimate channel bank or gully erosion by measuring the aerial differences between the sets of photos. The volume of bank erosion can be estimated for long term erosion and bank migration rates or for single hydrologic events, if photos are available before and after the event.

Observations should be made of the flow conditions near the site during low and high flow conditions. These observations could be useful in defining the flow path of floating debris and the flow patterns near the site. For an existing structure, the observations could be useful in defining the region where the flow is affected by the structure, i.e., contraction and expansion lengths. Also, any direct and indirect evidence related to the delivery potential of floating debris as mentioned in Chapter 3 of this manual should be noted. For example, abundant floating debris stored in the channel is direct evidence for high delivery potential. There is considerable direct and indirect evidence of debris generation that can be collected and used to evaluate the potential for debris accumulation at a site.

Measured sediment and discharge data are useful in estimating sediment yields of fine and coarse detritus. Unfortunately, this type of data is seldom available. Therefore, hydrological and sedimentation analyses are required to estimate the quantities of this type of debris. The only additional types of data needed for these analyses that have not already been mentioned are sediment gradation curves of the streambed and banks.

Among the factors to be considered are possible future changes in the type and quantity of debris that might result from changes in land use within the drainage basin. Some of these potential changes include floods, fires, urbanization, logging, grazing, agriculture, channel improvements, and conservation practices. As an example, logging in a previously virgin area could increase the quantity of floating debris introduced into the stream and change the nature of the debris problem from one of "medium floating" to "large floating" debris.

Lastly, any data needed for the debris structure should be obtained during the field investigations. An example would be the maximum allowable headwater and embankment height for culvert or debris-control countermeasure. This information might also be necessary in selecting the type of debris-control countermeasure best suited to the particular problem.

In summary, the following information should be obtained during the field investigations:

- Classification (as to the type) of the debris transported to the site.

- Information for estimating the quantity of debris.

- Land use and soil maps.

- Existing and historical survey data.

- Stream and watershed characteristics upstream of the site.

- Aerial photographs.

- Observations of the flow conditions near the site during low and high flow conditions.

- Direct and indirect evidence related to the delivery potential of floating debris.

- Information about future changes that could influence the quantity of debris.

- Data required for design, i.e., maximum allowable headwater elevation for a culvert structure.

6.1.2   Selecting the Type of Countermeasures

As noted in the previous chapter of this manual, there are a wide variety of countermeasures available for debris-control. A debris-control countermeasures matrix, presented in Table 6.1,

Table 6.1. Debris-Control Countermeasures Matrix.

COUNTERMEASURE	Floating Debris — Light	Medium	Large	Flowing Debris — Fine Detritus	Coarse Detritus	Bed Material — Boulders	MAINTENANCE — Estimated Allocation of Resources (H = High, M = Moderate, L = Low)	AESTHETICS[4]	ENV. IMPACT[5]	INSTALLATION EXPERIENCE BY STATE	DESIGN GUIDELINE[6]
GROUP 1. STRUCTURAL COUNTERMEASURE											
GROUP 1.A. CULVERT STRUCTURES											
Deflectors		x	x			x	H	A	L	CA	6.2.1
Racks	x	x					H	A	L	CT, CA	6.2.2
Risers				x	x		L	A	L	CA	6.2.3
Cribs	x				x		M	A	L	CA	6.2.4
Fin			x	x			M	A	L	SD, TN, CA	6.2.5
Dams and Basins				x	x		H	A	H	Widely Used	6.2.6
GROUP 1.B. BRIDGE STRUCTURES											
Deflectors		x	x			x	H – M	U	L	CA, MS, IN, OR, WI, LA	6.3.1
Fins		x	x				M	A	L	CA	6.3.2
Crib Structure		x	x				L	U	L	MS, CA, KS	6.3.3
R.T.S. – Iowa Vanes				x			M	A	M	IA	HEC-23[35]
R.T.S. – Permeable Spurs				x	x		M	A	M	AZ, CA, IA, MS, NE, OK, SD, TX	HEC-23[35]
R.T.S. – Impermeable Spurs				x	x		H	A	H	Widely Used	HEC-23[35]
In-channel Debris Basins				x	x		H	A	H	Widely Used	-
Flood Relief Sections		x	x				L	A	L	Widely Used	6.3.6
Debris Sweeper (Bridgeshark)		x	x				L	A	L	OK, VA, TN, OR	6.3.7
Booms	x	x					L	U	M	ID	-
D.F. – Freeboard		x	x				L	D	L	Widely Used	6.3.8
D.F. – Pier Type, Location, and Spacing		x	x				L	D	L	Widely Used	6.3.8
D.F. – Special Superstructure		x	x				L	D	L	TX, MS	6.3.8
GROUP 2. NON-STRUCTURAL COUNTERMEASURE											
Emergency and Annual Maintenance	x	x	x	x			H	U	M	Widely Used	-
Debris Management Plan	x	x	x	x			H	D	L	Widely Used	6.4

Notes:
1. "x" corresponds to a suitable device.
2. R.T.S. corresponds to River Training Structures.
3. D.F. corresponds to Design Features.
4. Classification for Aesthetics is: (1) U for Undesirable, (2) A for Acceptable, and (3) D for Desirable.
5. Classification for Environmental Impact is: (1) L for Low, (2) M for Medium, and (3) H for High.
6. Reference made above for the Design Guidance is related to the section indicated in this manual, i.e., information on deflectors for culverts is provided in Section 6.2.1.

has been developed to provide guidance in the selection of countermeasures suitable for various types of the debris. The matrix is organized to highlight the various groups of countermeasures and to identify their individual characteristics. The left side of the matrix lists the type of countermeasures available for the three general groups of structural countermeasures for culverts and bridges, and non-structural countermeasures for both. In each row of the matrix, the countermeasures suitable for the various types of debris are identified. The matrix also identifies the States that have used the countermeasure and the general level of maintenance resources required for the countermeasure. Finally, a resource for design guidelines is noted, where available.

In addition to the information contained within the matrix, the selection of the countermeasure should be based on the construction and maintenance costs, risk of failure, risk of property damage, and environmental and aesthetic considerations. The safety of highway traffic should also be considered in the selection of the countermeasure. The culvert end and the countermeasure should be located beyond the usual recovery area for errant vehicles or the countermeasure should be designed to enhance the drivers' chance of recovery. At existing sites where modifications cannot be made to meet this objective, an appropriate vehicle restraining device or an impact attenuating device should be provided on the roadside.

The countermeasure matrix (Table 6.1) was developed to identify distinctive characteristics for each type of countermeasure. Four categories of countermeasure characteristics were defined to aid in the selection and implementation of the countermeasures:

- Debris Classification
- Maintenance
- Installation/Experience by State
- Design Guideline References

These categories were used to answer the following questions:

- For what type of debris is the countermeasure applicable?
- What level of resources will need to be allocated for maintenance of the countermeasure?
- What States or regions in the U.S. have experience with this countermeasure?
- Where do I obtain design guidance reference material?

The Debris Classification Category describes the type of debris for which a given countermeasure is best suited or under which there would be a reasonable expectation of success. Conversely, this category could indicate the type of debris under which experience has shown that a countermeasure may not perform well or was not intended.

The type of debris considered for this category is based on the debris classification provided in Chapter 2 of this manual:

- Light Floating Debris
- Medium Floating Debris
- Large Floating Debris
- Flowing Debris
- Fine Detritus
- Coarse Detritus
- Boulders

The suitable countermeasure for each debris classification is indicated by "x". When the debris is comprised of more than one type of debris, the information provided in this category can be used as guidance in selecting a combination of countermeasures to address the debris problems.

The Maintenance Category identifies the estimated level of maintenance that may need to be allocated to service the countermeasure. The rating for this category is subjective, and it ranges from "Low" to "High." The ratings represent the relative amount of resources required for maintenance with respect to other countermeasures provided within the matrix (Table 6.1). A low rating indicates that the countermeasure is relatively maintenance free; a moderate rating indicates that some maintenance is required; and a high rating indicates that the countermeasure requires more maintenance than most of the countermeasures in the matrix.

The Aesthetics Category identifies the estimated level of appearance associated with the countermeasure with respect to other countermeasures provided within the matrix (Table 6.1). The rating for this category is subjective, and it ranges from "Undesirable" to "Desirable." An undesirable rating indicates that the countermeasure is noticeably unpleasing to the sight; an acceptable rating indicates that majority of the structure is pleasing to the sight; and a desired rating indicates that the countermeasure is noticeably pleasing to the sight.

The Environmental Impact Category identifies the estimated level of impact the countermeasure would have on the environment with respect to other countermeasures provided within the matrix (Table 6.1). The rating for this category is also subjective, and it ranges from "Low" to "High." A low rating indicates that the countermeasure does not adversely impact the environment or the impacts are considered short term; a moderate rating indicates that some adverse impacts could occur with implementation of the countermeasure; and a high rating indicates that the countermeasure would adversely impact on the environment.

The Installation/Experience by State category identifies DOTs that have used the countermeasure. This information was obtained from three sources: response of the DOTs to a debris-related questionnaire documented in "Debris Problems in the River Environment" (1979)[14]; Brice and Blodgett, "Countermeasures for Hydraulic Problems at Bridges, Volumes 1 and 2" (1978)[12, 13]; and correspondence between FHWA and DOT staff. **It is expected that additional information on state use will be obtained as this matrix is distributed and**

revised. Countermeasures that have been used by many States are given a listing of "Widely Used." The listing reflects both successful and unsuccessful experiences.

The countermeasures matrix (Table 6.1) is a convenient reference guide on a wide range of countermeasures applicable to addressing debris problems at culvert and bridge structures. A comprehensive plan of action would be to provide conceptual design and cost information on several alternative countermeasures, with a recommended alternative being selected based on a variety of engineering, environmental, and cost factors. The countermeasures matrix is a good way to begin identifying and prioritizing possible alternatives. The information provided in the matrix related to the suitable applications for the various types of debris and maintenance issues should facilitate preliminary selection of feasible alternatives prior to more detailed investigation.

6.1.3 Design for Bridges versus Culverts

The countermeasures provided in Table 6.1 have been divided into two groups: Structural countermeasures and Non-structural countermeasures. As seen in the table, the structural countermeasures have been further divided into structures available for culvert and bridge structures. Culverts, as distinguished from bridges, are usually covered with embankment and have structural material around the entire perimeter, although some are supported on spread footings with the streambed serving as the bottom of the culvert. Culverts are typically provided for small drainage basins and watercourses, and they are considered minor structures compared to bridges. Bridges, on the other hand, are usually used where the discharge of the watercourse is significant (larger drainage basins) or where the stream to be crossed is large in extent.

Because of the significant difference in size and function of these structures, some of the countermeasures available for culvert structures cannot be used for bridge structures, and vice a versa. Some countermeasures available for culvert structures are also available for bridge structures even though the intended purpose is different for the two structures. The debris deflector is an example of such a countermeasure. Debris deflectors are used at culverts to prevent debris from going through the culvert by deflecting it to the side of the structure where it is stored (debris retention), whereas debris deflectors at bridges are used to deflect the debris away from the pier and through the bridge opening (debris passage).

At many locations, either a culvert or bridge structure will satisfy both the structural and hydraulic requirements of the stream crossing. Structure choice at these locations should be based on construction and maintenance costs, risk of failure, risk of property damage, traffic safety, environmental, and aesthetic considerations. An additional deciding factor at these locations may be related to debris passage. Instead of providing a debris-control countermeasure at a proposed site, it may be desired to design the structure for debris passage. However, there are some obvious limitations in the case of culverts. There is no real assurance that doubling the size of a culvert will eliminate the threat of the culvert becoming plugged if debris poses a problem at the site. It is obvious that the probability of this occurring does decrease to some degree with increases in the size of the culvert. However, it is extremely difficult to demonstrate what level of protection would be obtained by such increases. Therefore, it may be necessary to use a bridge structure that is designed for debris passage with a higher degree of certainty in lieu of a culvert structure even though it can adequately convey the anticipated flows.

Both types of non-structural countermeasures can be considered for culvert and bridge structures. However, a debris management plan would more likely be implemented at a bridge than a culvert due to the high cost and allocation of resources required to develop and execute such a plan making it infeasible for small drainage structures.

6.1.4   Existing Structures Versus Proposed Structures

The selection and design of the countermeasures presented in Table 6.1 could depend on if the countermeasure is for an existing or proposed structure. Constraints at an existing structure can prevent the use of certain countermeasures or influence the design of the countermeasure. The constraints could be related to the physical conditions at the site; the structure itself; monetary reasons; environmental or maintenance requirements; limited or no access to the culvert or bridge; or other reasons. Recent development adjacent to the watercourse upstream of an existing structure might prevent the use of an in-channel debris basin or dam because of potential flooding impacts or political pressure from the residents. The geometry of the existing structure could influence the configuration and dimensions of the proposed countermeasure. Certain countermeasures might require that part of the existing structure be demolished and significantly modified, making it too expensive to implement. It is also possible that environmental restrictions due to fish passage or vegetation removal could limit the type of countermeasure that could be selected and constructed at a particular site.

All of the countermeasures presented in Table 6.1 could be used at proposed structures. However, this is not the case for existing structures. Unfortunately, the most common type of countermeasures used for bridge structures are usually infeasible to implement at existing bridge structures. These measures are identified in Table 6.1 as design features, i.e., "D.F.", and they include adequate freeboard, the use of special superstructure, and considerations to the type, location, and spacing of piers for reducing the potential of accumulation. These countermeasures can easily be incorporated into the design of a proposed structure, whereas they are difficult to implement at an existing structure. The anticipated debris accumulation on a proposed bridge can be considered in the: (1) design of the hydraulic opening through the bridge to safely convey the design flood without overtopping the structure, (2) structural design of the bridge components to withstand the increase in lateral and overturning forces associated with the debris accumulations, and (3) design of the pier and abutment foundations to prevent undermining of the structure by the significant scour associated with debris accumulation.

Proposed structures also have the benefit over existing structures in that access for maintenance can be included in the design of the structure. Access to a proposed structure can be incorporated into the design of the highway embankment where this might not be a viable option for an existing highway embankment.

For existing structures, the problems associated with debris are usually more easily understood. There is generally sufficient information on the type of debris, quantity of debris transported to the site, and the associated problems with the debris available to select and design countermeasures to address the problem. In some instances where the investment is relatively small and there is little chance of interruption to current operations, it may be more desirable at a proposed site to select and design the countermeasure after the problems with debris have developed and are fully understood.

6.1.5 Maintenance Accessibility of the Countermeasure

Maintenance is an important factor to consider in selecting a debris-control countermeasure or designing a bridge structure. This should entail both regular and emergency maintenance activities. Considerations should be made as to the ease and cost of maintaining the countermeasure and accessibility to the countermeasure for performing the maintenance work. A countermeasure that is more expensive to construct may be more desirable if it is easier and less expensive to maintain.

Provisions should be made for access to the countermeasure for maintenance purposes. Maintenance personnel should be consulted with when designing the access to the site. Unfortunately, access is often difficult to provide, and it may not be provided for countermeasures designed for secondary highways or lower class roads. If access roads to the countermeasure are impractical and the risk associated with flooding is high, it may be necessary to provide an area near the countermeasure where mechanical equipment, such as a crane, could perform maintenance activities, such as, debris and sediment removal.

Maintenance accessibility for debris removal should be considered in the design of a new or replacement bridge. There are certain features that can be incorporated into the design to simplify debris removal. For instance, the use of solid wall piers that extend slightly upstream of the edge of the bridge deck. This type of pier configuration provides for easier removal of debris than other pier types. Debris not only accumulates more readily on multiple-column piers, but also may become entangled between the columns along the full width of the underside of the bridge, making it extremely difficult to remove the debris and/or causing access problems for the debris-removal crew. Debris trapped on trusses and piers with multiple columns can be entangled among multiple structural elements. The entanglement makes debris removal more difficult and increases the possibility that the bridge could be damaged during the removal operations. Hammerhead piers are an alternative to multiple columns. This type of pier eliminates the potential for entanglement. However, debris removal is still difficult since the pier nose is well beneath the bridge deck and it is extremely difficult to access and lift the debris from the bridge deck. Superstructures that allow access to the pier nose from directly above also ease debris removal. A wide deck with a simple parapet and adequate load-bearing capacity for heavy equipment at the upstream edge provides the best opportunity for debris removal from the bridge deck.

Access should be provided to the substructure of bridges to ensure prompt and complete removal of debris that accumulates on the structure. Debris removal can usually be accomplished during low flows with tracked vehicles for bridges over small streams. However for bridges over large rivers, a barge might be required to remove the debris, so a launching site for the barge may be necessary at such a site.

6.2   DESIGN GUIDELINES FOR CULVERTS

6.2.1   Debris Deflectors for Culverts

1.   FUNCTION – The function of a debris deflector (Figure 5.1 through Figure 5.9) is to divert medium and large floating debris and large rocks from the culvert inlet for accumulation in a storage area where it can be removed after the flood subsides. Their structural stability and orientation with the flow make deflectors particularly suitable for large culverts, high velocity flow, and debris consisting of heavy logs, stumps, or large boulders.

2.   STORAGE AREA – The storage area provided must be adequate to retain the anticipated type and quantity of debris expected to accumulate during any one storm or between cleanouts.

3.   TYPE OF MATERIAL – Debris deflectors are usually built of heavy rail or steel sections (Figure 5.1 through Figure 5.9), although timber (Figure 5.8 and Figure 5.9) and steel pipe are sometimes used when the type of debris consists of light floating debris and/or fine detritus. The decision to use timber in lieu of steel could also be based on the availability of the material within the region and construction costs. Wire and post debris deflectors (Figure 5.3) have also been used for light floating debris. For economy, salvaged railroad rails may be used if available. Figure 5.3 shows a deflector that uses a cable as its lower longitudinal member. This modification has proved to be superior in locations where heavy boulders damage rigid members.

4.   LOCATION AND ORIENTATION – The deflector should be built at the culvert entrance and aligned with the stream rather than the culvert so that the accumulated debris will not block the channel. Individual deflectors can be built over each pipe (Figure 5.4) or a single deflector can be built over multiple pipe culverts (Figure 5.5). The deflector may be placed at the culvert entrance or a distance of 1 culvert dimension upstream. The apex of the deflector will "point" upstream.

5.   DIMENSIONS – The angle at the apex of the deflector should be between 15° and 25°, and the total area of the two sides of the deflector should be at least 10 times the cross-sectional area of the culvert. The base width and height of the deflector should be at least 1.1 times the respective dimensions of the culvert. The upstream member is vertical on most installations. However, a sloping member at the apex (sloping downstream from bottom of member) would reduce the impact of large floating debris and boulders, and probably prevent debris from gathering at that point. Therefore, deflectors with a sloping member at the apex are recommended over a vertical upstream member.

6.   BAR SPACING – Spacing between vertical members should not be greater than the minimum culvert span dimension nor less than 1/2 the minimum dimension. A spacing of 2/3 the minimum dimension is commonly used. In addition to what is required for structural support, spacing of the horizontal bars along the sides of the deflector follow similar characteristics. Where headwater from the design flood is expected to be above the top elevation of the deflector and floating debris is anticipated, horizontal members should be placed across the top. The spacing of horizontal members on the top should be no greater than 1/2 the smallest dimension of the culvert opening.

6.2.1.1   Debris Deflector Example (Culvert)

Given:

Circular Culvert, Diameter = 1.8 m
Sediment material comprised of coarse detritus and medium floating debris

Determine:

Determine the dimensions of a triangular shaped debris deflector.

Solution:

Step 1: Determine the height and minimum width of the debris deflector

Height, H = 1.1(Culvert diameter, D) = 1.1(1.8) = 1.98 m (use 2.0 m)

Minimum width, W_{min} = 1.1D = 1.1(1.8) = 1.98 m (use 2.0 m)

Step 2: Select desired apex angle, bar spacing and thickness

Apex angle, α, can range from 15° to 25°, use 20°

Bar spacing, s, can range from 1/2(D) to D. 2/3(D) is common spacing, so use 1.2 m.

Deflector will be constructed out of 76-millimeters-thick, t, steel rails. The thickness of the rail was selected taking into account the type of debris, availability of the material, cost, and structural stability.

Step 3: Determine the side length of the debris deflector

A trial and success procedure is required to determine side length of the debris deflector:

a.   determine the minimum side area of deflector,
b.   assume a length of the deflector,
c.   determine the number of vertical and horizontal bars,

d. compute the gross area of the deflector,
e. compute the total area of the steel rails,
f. compute the net area of the deflector, and
g. complete substeps "b" through "f" until the net area of the deflector is slightly greater than the minimum side area of the deflector.

Implementing these substeps "a" through "g"

Substep "a"
Minimum side area of deflector on each side = 5(Area of culvert)
Minimum side area of deflector on each side = $5(\pi D^2/4) = (5)((3.14)(1.8)^2)/4) = 12.72$ m^2

As this is a trial and success solution, this example will only show the values in the first and final iteration. The "final" iteration will be a bold, italicized value provided within parenthesis.

Substep "b"
Assume a length, L, of 6.7 m   *(Final: **7.9** m)*

Substep "c"
Number of vertical bars = (L + s)/(s + t)
                        = (6.7 + 1.2)/(1.2 + 0.076) = 6.19 -> use 7   *(Final: **8** bars)*
Number of horizontal bars = (H + s)/(s + t)
                          = (2.0 + 1.2)/(1.2 + 0.076) = 2.51 -> use 3   *(Final: **3** bars)*

Substep "d"
Gross area per side = (L)(H) = (6.7)(2.0) = 13.4 m^2 *(Final: **15.80** m^2)*

Substep "e"
Area of bars = (number of horizontal bars)(t)(L) + (number of vertical bars)(t)(H)
Area of bars = (3)(0.076)(6.7) + (7)(0.076)(2.0) = 2.59 m^2 *(Final: **3.02** m^2)*

Substep "f"
Net Area = Gross area – Area of bars = 13.40 – 2.59 = 10.81 m^2 *(Final: **12.78** m^2)*

Substep "g"
Compare Net Area and Side Area: 10.81 m^2 less than 12.72 m^2. Therefore increase length (substep "b") and try another iteration.

Step 4: Determine the distance to the apex of the deflector and the width of deflector

Distance to the apex of the deflector = (L)(cos(0.5α)) = (7.9)(cos(10)) = 7.78 m
Width of the deflector = (2)(L)(sin(0.5α)) = (2)(7.9)(sin(10)) = 2.74 m

Figure 6.1 depicts the resulting culvert debris deflector developed in the example. The figure is for illustration purposes only and should not be used as a specification or detail.

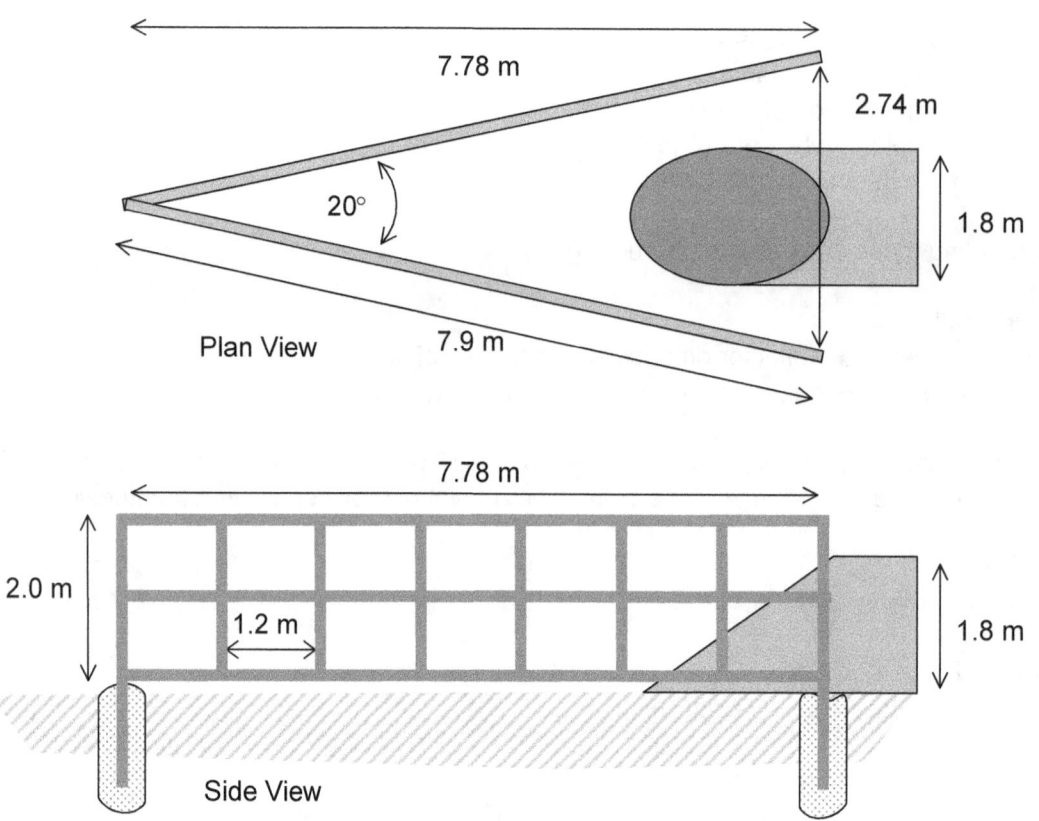

Figure 6.1. Debris deflector designed in example.

6.2.2 Debris Racks for Culverts

1. FUNCTION – The function of a debris rack (Figure 5.10 through Figure 5.22) is to essentially create a barrier across the stream channel to trap light and medium floating debris that is too large to pass through the culvert.

2. STORAGE AREA – The storage area provided must be adequate to retain the anticipated type and quantity of debris expected to be accumulated during any one storm or between cleanouts. If a large debris storage area is provided upstream of the rack location, the frequency of maintenance can significantly be reduced and added safety is provided against overtopping of the installation during a single storm.

3. TYPE OF MATERIAL – Debris racks are usually built of heavy rail or steel sections, although they can be constructed out of various types of material. Inclined racks and rubber tires have been used to help reduce the impact of heavy debris striking at high velocity. Chain-link fence has also been used to remove light floating debris from low velocity streams. This type of barrier is particularly advantageous in tidal areas where the function of flap or check

gates can be hampered by light debris collecting on the gate seats and thereby blocking complete closure of the gates. Since vertical racks receive the full impact of floating debris and boulders, their structural design should incorporate brace members set in concrete.

4. LOCATION AND ORIENTATION – Debris racks may be vertical or inclined and may be placed over the culvert inlet (Figure 5.10, Figure 5.14, Figure 5.15, Figure 5.21, and Figure 5.22) or upstream from the culvert (Figure 5.11, Figure 5.12, Figure 5.13, Figure 5.16, Figure 5.17, Figure 5.18, Figure 5.19, and Figure 5.20). Racks should not be placed in the plane of the culvert entrance, since they become easily plugged. Where a well-defined channel exists upstream of the culvert, the debris rack should be placed upstream from the culvert entrance a minimum distance of two times the culvert diameter. However, it should not be placed so far upstream that debris can enter the channel between the structure and the culvert entrance, or accessibility to and maintenance of the structure becomes difficult and/or costly. In addition, right-of-way constraints are important considerations in locating debris racks. Debris racks generally do not have top or horizontal members that extend from the rack to the culvert headwall, although there are exceptions.

5. DIMENSIONS – The total straining area of a rack should be at least ten times the cross-sectional area of the culvert being protected. The overall dimensions of the rack should be a function of the amount of debris expected per storm, the frequency of storms, and the schedule of expected cleanouts. When a rack is installed at the upstream end of the wingwalls, it should be at least as high as the culvert. Also, the height of racks should allow some freeboard above the expected depth of flow in the upstream channel for the design flood. Racks 10 to 20 feet high have been constructed.

6. BAR SPACING – Spacing between vertical members should not be greater than 2/3 the minimum culvert dimension nor less than 1/2 the minimum culvert dimension. This spacing permits the lighter debris to pass through the rack and the culvert. In urban areas, bar spacing of racks should be a maximum of 6 inches and tied to the culvert headwall by top bars to prevent children from entering the culvert. Unfortunately, the close spacing of the bars creates a debris trap and increases the maintenance required. To reduce the amount of debris becoming trapped, it is preferable to have the lowest edge of rack about six inches above the flow line of the ditch, permitting some debris to pass under the rack during low flows.

6.2.2.1 Debris Rack Example (Culvert)

Given:

>Culvert Diameter = 1.8 m
>Design Discharge = 9.9 m³/s
>Design Headwater Depth = 2.7 m
>Upstream channel width = 7.3 m
>Flow carries light to medium debris
>Rack constructed out of 76-millimeter-thick steel rails

Determine:

Determine the dimensions of a vertical debris rack comprised of two horizontal members.

Solution:

Step 1: Determine the minimum area of the debris rack

Minimum area of debris rack, A_{rack} = 10(Area of culvert)
Minimum area of debris rack, A_{rack} = 10($\pi D^2/4$) = 10(3.14)(1.8)^2)/4 = 25.45 m^2

Step 2: Determine the number of vertical bars and spacing

Minimum spacing = 1/2(D) = s_{min} = 1/2(1.8) = 0.9 m

Maximum spacing = 2/3(D) = s_{max} = 2/3(1.8) = 1.2 m

Number of vertical bars for minimum spacing = $(w - s_{min}) / (s_{min} + t)$ =
$$= (7.3 - 0.9) / (0.9 + 0.076) = 6.6 \rightarrow \text{use } 7$$

Number of vertical bars for maximum spacing = $(w - s_{max}) / (s_{max} + t)$ =
$$= (7.3 - 1.2) / (1.2 + 0.076) = 4.8 \rightarrow \text{use } 5$$

Try 7 vertical bars for debris rack (assume this is the smallest material and fabrication cost).

$$\text{Spacing} = \frac{(w - t(\text{number of vertical bars}))}{(\text{number of vertical bars} + 1)} = \frac{(7.3 - 0.076(5))}{(5 + 1)} = 1.15 \text{ m}$$

The spacing of 1.15 m falls between the minimum (0.9 m) and maximum (1.2) spacing values. Therefore the spacing is adequate.

Step 3: Determine the height of the debris rack

$$\text{Approximate height, } H = \frac{A_{rack}}{(n_v + 1)(s)} + (t)(\text{number of horizontal bars}) = \frac{25.45}{(6)(1.15)} + 0.076(2) = 3.84 \text{ m}$$

(note: variable "n" is the number of vertical bars in the rack – in this example "5").

Use a height of 4 m to account for additional loss in area from the horizontal bars. No adjustment required to satisfy freeboard since the design headwater elevation is about 1.3 m below the top of the rack.

Gross area = (w)(H) = (7.3)(4) = 29.2 m^2

Area of bars = (number of horizontal bars)(t)(w) + (number of vertical bars)(t)(H)
Area of bars = (2)(0.076)(7.3) + (5)(0.076)(4) = 2.63 m^2

Net Area = Gross area – Area of bars = 29.2 – 2.63 = 26.57 m^2

Compare Net Area and Minimum Rack Area: 26.57 m^2 is greater than 25.45 m^2. Therefore, the design is adequate.

6.2.3 Debris Risers for Culverts

1. FUNCTION – Debris risers (Figure 5.23 through Figure 5.25, and Figure 5.32) generally consist of vertical culvert pipes that are commonly used as relief structures, either independent of the main culvert or in conjunction with it. Risers are used where considerable height of embankment is available and the type of debris consists of flowing masses of clay, silt, sand, sticks, or medium floating debris without boulders. They are seldom structurally stable under high-velocity flow conditions because of their vulnerability to damage by impact, and they are usually suitable for culvert installations of less than 54-inch diameter.

2. STORAGE AREA – Storage area must be provided to adequately retain the anticipated type and quantity of debris expected to be accumulated during any one storm or between cleanouts. The use of risers induces deposition of the sediment material upstream of the riser as a result of the ponding created by the riser.

3. TYPE OF MATERIAL – Debris risers are usually built out of corrugated metal pipe, although they can be constructed out of steel pipe. A corrugated metal pipe reducing elbow can be used to connect risers to an existing culvert inlet, although damage to the metal elbow from falling rocks may occur. Occasionally, concrete is placed inside the elbow to prevent the metal from wearing through by this abrasive action. A solution for extremely severe conditions is to connect riser and culvert by a concrete junction box having the inside shaped as an elbow.

4. LOCATION AND ORIENTATION – Debris risers are typically vertical, however they have been built at an angle between vertical and the stream grade. This reduces the impact of debris at the elbow and assists in moving debris through the culvert. Risers can be either an independent structure or connected to an existing culvert. If connected to the existing culvert, the riser should be located a minimum of 3 feet or one half of the culvert diameter, whichever is greater, from the existing culvert headwall.

5. DIMENSIONS – Good practice will build riser pipes at least 36 inches in diameter to provide an area large enough for maintenance access. To avoid vibration of the riser pipe and unstable flow conditions, the riser diameter should be about 1 foot larger than the culvert diameter. If the embankment is of sufficient height, provisions should be made to extend the riser vertically, if necessary. In the case of corrugated metal pipe risers, this can be accomplished by means of standard coupling bands.

6. GRATE AND SLOT FEATURES – Debris risers should be covered by a grate or cage to prevent clogging of the culvert. The grate bars can be reinforcing steel or other such material with vertical spacing no greater than 1/2 the culvert diameter. Slots or holes are placed in the sides of the riser to carry low flow. It is preferable to have these holes punched before galvanizing to avoid deterioration by rust. The holes are considered to have no hydraulic capacity under peak flow conditions because of the likelihood of their becoming plugged by light floating debris and silt. It is also desirable to connect the grate bars to a coupling band, rather than directly to the riser pipe, so the grate can be removed should cleaning be required.

6.2.4 Debris Cribs for Culverts

1. FUNCTION – A debris crib (Figure 5.26 through Figure 5.28) is particularly adapted to small-size culverts where a sharp change in stream grade or constriction of the channel causes deposition of coarse detritus at the culvert inlet. Debris can almost envelop a crib without completely blocking the flow and plugging the culvert. Cribs are somewhat similar to risers, however cribs are more appropriate than risers where the culvert has little cover and the debris is comprised of coarse detritus. Due to the debris type and site conditions associated with debris cribs and also risers, these two types of countermeasures have shown to be the most consistently successful in producing an efficient, maintenance-free installation for culverts.

2. TYPE OF MATERIAL – Debris cribs are usually constructed of precast concrete or wood, and precast concrete should be used when the debris consists of medium to large cobbles.

3. LOCATION AND ORIENTATION – The crib is usually placed directly over the culvert inlet and is generally built in log-cabin fashion although other designs have been used. Debris cribs may be open (Figure 5.26 and Figure 5.27) or covered (Figure 5.28).

4. DIMENSIONS – The spacing between the bars should be about 6 inches. This spacing also applies to the horizontal top members of a covered crib. The height of an open crib should be higher than the depth of debris deposited at the structure. When an open crib is used as a riser and an accumulation of detritus is expected to build up, provision should be made for the height of the structure to be increased as needed (Figure 5.26 and Figure 5.27). Cribs have been built as high as 50 feet above a pipe invert with little change in the efficiency of the facility.

6.2.5 Debris Fins for Culverts

1. FUNCTION – Debris fins are thin walls installed upstream of the culvert parallel with the flow (Figure 5.29 through Figure 5.35). They have been used successfully with large culverts or for multiple box culverts where the debris consists mostly of floating material that can pass through the culvert if oriented parallel with the culvert barrel. Debris that is not aligned by the fin to pass through the culvert is retained at the front of the fin for later removal by maintenance personnel. If the fin is sloped upward toward the culvert, the debris that does not

pass through the culvert can float upward and prevent debris from blocking the culvert inlet. Fins are generally not used on culverts with a minimum dimension less than 4 feet.

2. TYPE OF MATERIAL – Debris fins are usually concrete, although they have been constructed of steel and timber.

3. LOCATION AND ORIENTATION – Debris fins can extend from the interior walls of culverts (Figure 5.29 through Figure 5.31) or located on the centerline of a single culvert (Figure 5.32 through Figure 5.34). The upstream end of the fin should be rounded and sloped upward toward the culvert (Figure 5.29 and Figure 5.30) to reduce impact, turbulence, and the probability of gathering debris, rather than vertical as shown in Figure 5.31 through Figure 5.34. If the upstream end of the fin is vertical, rounding that edge would be preferable to a square edge (Figure 5.31).

4. DIMENSIONS – A debris fin is usually constructed to the height of the culvert; thus, its effectiveness is limited after the inlet becomes submerged. Field experience indicates the fin length should be 1 1/2 to 2 times the culvert height. The leading edge would thus have a slope from 1-1/2:1 to 2:1 (from 33.7 to 26.6 degrees). The thickness of the fin should be the minimum needed to satisfy structural requirements in order to minimize disturbance to the flow.

6.2.6 Debris Dams / Basins for Culverts

1. FUNCTION – Debris dams are structures placed across a well-defined channel to form a barrier that impedes the stream flow. The dams also form a basin that provides storage for deposits of detritus and floating debris (Figure 5.36 through Figure 5.39). Debris dams and basins are used at sites that convey heavy debris loads where it is economically impracticable to provide a culvert large enough to convey the surges of debris. They are also used to trap heavy boulders or coarse gravel that would clog culverts, especially on low fills. In some locations, debris dams have been built to provide the added benefit of ground water recharge resulting from ponded water.

2. TYPE OF MATERIAL – Debris dams are usually earth or rock filled structures. Debris dams, however, can be built out of metal (Figure 5.36), rock held in place by wire (Figure 5.37), (i.e., gabions), or precast concrete beams placed in crisscross or log-cabin fashion with rock dumped between the members (Figure 5.38 and Figure 5.39).

3. LOCATION AND ORIENTATION – Debris dams and basins are usually placed some distance upstream from a culvert. However at some locations, the highway embankment can serve as the embankment for the debris dam.

4. DESIGN FEATURES –There are several features that must be considered in the design of the debris dam and basin. Some of the important features are the embankment, inlet protection, outlet structure, and emergency spillway structure. Information on the design of these features and sedimentation basins in general is provided in "Design of Sedimentation Basins".[22] Prior to initiating the design of the debris dam, state agencies should be contacted

to ensure that state regulations are met in the design of the structure. Also, hazards created by the failure of the debris dam need to be considered and evaluated during the design of the structure.

Various items must be considered in the design of the embankment. One of the more important items is the height of the embankment. The top of the embankment should be set at an elevation sufficiently above the maximum ponding elevation associated with the design volume of runoff and debris to assure that the runoff and debris are contained with a high level of certainty, i.e., embankment has adequate freeboard. When defining this elevation, the debris storage volume should be based on the assumption that the deposition slope of the debris is horizontal and not one half of the natural valley slope that has commonly been used. This assumption eliminates the potential of the embankment being overtopped due to the momentum of the flow, which has occurred for some of the debris dams and basins in the Los Angeles area designed assuming a deposition slope equal to one half of the natural valley slope. Stability of the embankment is also a major concern. The embankment should be designed to withstand the total forces from soil and hydrostatic pressure, seepage uplift, and earthquake on the structure. Special considerations for slope stability should be made for earth-fill embankments. The upstream and downstream slopes of the earthen structure depends on the soil material used to construct the embankment; however, the slope typically ranges between 2.5 to 1 and 3 to 1 for the upstream face and between 2 to 1 and 3 to 1 for the downstream face. Another important item for earthen structures is slope protection. Both the upstream and downstream face of the embankment should be protected with some type of slope protection measures, i.e., vegetation, riprap, matting, or mulch, to prevent erosion of the embankment.

Occasionally, extensive excavation below the natural streambed is necessary to provide the required storage for the debris. For this type of basin design, the upper end of the basin should be protected with revetment to prevent any upstream erosion of the streambed due to headcutting.

An outlet structure should be provided to drain the floodwater temporarily stored behind the structure. The structure could be either a closed conduit consisting of a culvert with a riser set above the expected level of the debris deposit or an open channel acting as a weir structure. The design of the structure will have an influence on the design volume of the basin and embankment height. In general, an outlet structure designed to convey more of the runoff volume will reduce the design volume of the basin and lower the embankment height, but the cost of the structure will increase. Therefore, several different types and sizes of the outlet structure should be considered in the design of the structure to optimize the total cost of the debris dam. Significant scour can develop downstream of the outlet structure due to the high velocity, turbulent flow leaving the structure and the significant reduction in the sediment load resulting from the upstream deposition. Therefore, protection measures must be provided at the downstream end of the structure to protect the structure and embankment from failure due to undermining. Access for maintenance and repair work should be provided to the upstream and downstream ends of the structure.

The debris dam must have an emergency spillway to safely convey flows greater than the design event. The spillway should be located off to one side of the embankment and excavated into an adjoining hillside since this location is more stable against breaching than a spillway over a fill or over the embankment structure. Outlet structures designed as open channels can also be designed to serve as the emergency spillway. Protection measures must be provided at the downstream end of the spillway to protect the structure and embankment from failure caused by significant scour.

6.2.7 <u>Combined Debris Controls for Culverts</u>

Each drainage basin presents its own debris problem. Often more than one problem exists at a site and two or more types of debris-control countermeasures are required to adequately address the problems. Combined measures can also be used at locations where it may be preferable to remove the larger debris at a location upstream from the culvert and to remove the smaller material nearer the culvert inlet. Combined measures can also be used at locations where it may be advisable to install two types of devices so that one will function if the other fails. An example of this is shown in Figure 5.24 where a debris riser was installed over the entrance of a culvert to assure water is conveyed through the culvert in the event that the culvert entrance becomes plugged. A photograph of this installation after a flood event is shown in Figure 5.25. Other examples of combined countermeasures are shown in Figure 5.32, Figure 5.34, and Figure 5.40. In these cases, Figure 5.32 and Figure 5.34 show a culvert protected by both a debris fin and a debris riser. Figure 5.40 shows an installation consisting of a debris dam and settling basin with a debris deflector at the inlet and a debris riser.

6.3 DESIGN GUIDELINES FOR BRIDGES

6.3.1 <u>Debris Deflectors for Bridges</u>

1. FUNCTION – Debris deflectors are placed upstream of bridge piers to divert and guide debris through the bridge opening. Deflectors are used where the debris consists of medium to large floating debris.

2. TYPE OF MATERIAL – Debris deflectors attached to the pier are usually constructed of steel rails, whereas steel piles filled with concrete are used for deflectors located some distance upstream.

3. LOCATION AND ORIENTATION – Debris deflectors can be attached to the pier or located at some distance upstream from the pier. The effectiveness of deflectors is largely controlled by the direction of stream flow. Changes in flow direction can cause the deflector to be ineffective and in some cases actually worsen the situation. Therefore, deflectors can be greatly improved if the flow direction in the stream can be stabilized by auxiliary structures such as guide banks which confine and stabilize the flow in a certain direction. The flow patterns around the deflector are complex and cannot be easily predicted. The effectiveness of the structure is difficult to assess. Therefore, in the determination of proper location and configuration of the deflector, physical modeling is encouraged to assure proper functioning of deflector for various discharges.

6.3.2 Debris Fins for Bridges

1. FUNCTION – Debris fins are thin walls installed upstream of the bridge parallel with the flow (Figure 5.35). Debris fins have been successfully used to align debris with the waterway opening and to avoid the accumulation of debris on bridge piers. They are used when the debris consists mostly of floating material. Fins have also been successful in reducing ice clogging by displacing ice sheets upward along the sloping top surface.

2. TYPE OF MATERIAL – Debris fins are usually concrete, although they have been constructed of steel and treated timber piling and bracing.

3. LOCATION AND ORIENTATION – Debris fins are usually located on the centerline of the bridge piers, and they should be carefully aligned with the flow in order to avoid increasing the projected pier width and a corresponding greater depth of pier scour. The upstream end of the fin should be rounded and sloped upward toward the bridge to reduce impact, turbulence, and the probability of gathering debris.

4. DIMENSIONS – The debris fin consists of a vertical and sloped section. The vertical section exists from the upstream face of the pier to 1.8 m (6 ft) upstream and has a minimum height of 0.76 m (2.5 ft) above the maximum water surface elevation for the design flood event. The sloped section has a height equal to the maximum water depth at the upstream end of the vertical section. The sloped section extends upstream a distance of twice the maximum water depth. The profile of the sloped section consists of a 3:1 sloped segment and a curved segment with a point of intersection located one-half of the maximum water depth above and downstream of the upstream end of the fin. The overall width of the debris fin transitions from the width of the bridge pier to a width of 0.3 m (1 ft) at the upstream end of the fin. The debris fin foundation must be sufficient to withstand the predicted scour depth.

6.3.3 Crib Structures for Bridges

1. FUNCTION – Debris cribs are used for open-pile bents to prevent debris from trapping and accumulating between the piles. The crib structure is constructed around the existing pier structures by doweling the sheathing members directly into the existing piers or to vertical columns that are tied into the foundation of the existing piers.

2. TYPE OF MATERIAL – Debris cribs are usually built of timber or metal sheathing, although concrete sheathing has been used.

3. SPACING – The effectiveness of debris cribs is largely dependent on the spacing between the sheathing members. Unfortunately, there are no guidelines available for defining the spacing of the crib structure. In general, large spacing should be avoided since it creates a favorable condition for entrapping and accumulating debris.

4. FLOW DIRECTION – Special considerations should be made when the pile bents are skewed to the approaching flow. The narrow openings created by the structure increase the potential for debris trapping, and debris that would normally pass through the pile bents could accumulate on the structure.

6.3.4 River Training Structures for Bridges

River training structures are structures placed in the river flow to create counter-rotating streamwise vortices in its wakes to modify the near-bed flow pattern to redistribute flow and sediment transport within the channel cross section. Design guidelines for this type of structure are provided in HEC-23[35] as Design Guideline 9. The guidelines provided in HEC-23 cover the longitudinal extent of spur field, spur length, spur orientation, spur permeability, spur height and crest profile, bed and bank contact, spur spacing, shape and size of spurs, and rock sizes.

6.3.5 In-Channel Debris Basin for Bridges

In-channel debris basins are structures placed across well-defined channels to form basins which impede the streamflow and provide storage space for deposits of detritus and floating debris. Unfortunately, no design guidelines exist for these types of structures. The flow patterns around these structures are complex and cannot be easily predicted. Therefore, a physical model should be used to design and analyze the structure to assure that it functions properly for various discharges.

In-channel debris basins for floating debris have been used in parts of Europe. Two such structures are the "Arzbach Treibholzfang" and the "Lainbach Treibholzfang". Both of these structures were designed from physical model testing conducted at the Hydraulics Laboratory of the Technical University of Munich[61]. The two structures have similar configurations, however there are some noticeable differences between them. The "Lainbach" structure was built with a double row of posts that was later found to be unnecessary, so the "Arzbach" structure was built only with a single row of posts. The posts within the two structures are the same. They have a diameter of 0.66 m (2.2 ft) and a height of 4 m (13.1 ft) above the channel bed. They are comprised of a steel sleeve with a concrete core with each post set into a concrete foundation that is supported on piles extending 4.4 m (14.4 ft) below the channel bed. Both of the structures have riprap revetment along the bed and side slopes of the channel upstream and downstream of the posts to protect against erosion. Another difference between the two structures is that a performed scour hole downstream of the post was incorporated into the design of the "Arzbach" structure. The maintenance requirements for these structures are high with debris having to be removed periodically and possibly on an annual basis.

6.3.6 Flood Relief Structures for Bridges

Flood relief structures are flow through or overtopping structures that divert excess flow and floating debris through the structure and away from the bridge structure. These structures can significantly reduce the risk of significant damage or failure of the structure by reducing the pressure of the flowing water on the increased width of a pier resulting from the lodged debris and the amount of debris conveyed to the bridge. These structures were determined to be very effective in preventing failure of several bridge structures with debris accumulations during severe flooding in Pennsylvania and New York from Hurricane Agnes[47]. Therefore, a flood relief structure should be considered for bridges that have a high potential for debris accumulation and where there is space available and no physical constraints that would otherwise preclude their use.

Flood relief structures should be located near the ends of the bridge. These structures can be incorporated into the design of a bridge, where the anticipated debris accumulation is included in the design of the structure, to function as an emergency structure for conveying flows greater than the design discharge. They can also be utilized at existing bridges where debris accumulated on the structure has significantly reduced the discharge conveyed through the bridge and has caused significant increases in the upstream water surface elevation.

The discharge that a relief bridge would need to convey can be estimated using the following procedure:

1. Compute the water surface profile through the bridge for the design discharge, assuming no debris accumulation on the structure.

2. Estimate the location and extent of the debris accumulation using the procedures provided in Chapter 3 of this manual.

3. Reflect the accumulated debris and re-compute the water surface profile through the bridge for the design discharge to determine the effect the debris accumulations has on the upstream water surface elevation.

4. Compute a rating curve of discharge versus upstream water surface elevation for the bridge structure with debris accumulations.

5. Determine the maximum allowable water surface elevation upstream of the bridge structure using topographic mapping, historical flood information, and information from the field investigation. This elevation could also be defined as the elevation associated with potential failure of the bridge caused by the increase in hydraulic loading on the structure due to the debris accumulation.

6. Determine the flow through the bridge structure for the maximum allowable water surface elevation using the rating curve computed in the fourth step.

7. Determine the design discharge for the relief structure by subtracting the discharge computed in the previous step from the design discharge.

Relief structures should be protected with revetment where significant damage to the structure is undesired or when the anticipated difference between the upstream and downstream water surface elevations is large and there is a potential of catastrophic flooding downstream of the structure. Revetment should also be provided for the downstream slopes of highway embankments that are designed for overflow, or that are subject to overtopping, and the anticipated drop between the upstream and downstream water surface elevations is large.

6.3.7 Debris Sweepers for Bridges

A debris sweeper is a device, generally made of polyethylene, which is attached to a vertical stainless steel cable or column affixed to the upstream side of the bridge pier. The debris sweeper travels vertically along the cable or column as the water surface rises and falls. The devices are also rotated by the flow, causing floating debris to be deflected away from the pier and through the bridge opening. Two devices could be placed on the same track with one of the devices being completely submerged while the other device is near the water surface. The devices could be aligned with the pier or offset from the pier, and special considerations on the placement of the devices are required for skewed flow conditions. If access to the substructure from the bridge deck is a problem, then a column application can be utilized or the devices can be installed using a boat. Debris sweepers can be used for most types of floating debris with the larger, heavier debris requiring a stronger bracket design. Several States are still assessing the use of such sweepers.

Maintenance and inspection of these devices is recommended after a high-water event. All cable and anchors for the bracketing system should be inspected for proper tension, and any debris near the device and/or bracket system should be removed immediately, so that the performance of the device is not compromised during subsequent events.

An important design guideline appears to be carefully checking the suitability of the site to the sweeper application. For example, the device would not be an appropriate measure if the design log length is greater than the effective opening between the piers.

6.3.8 Design Features for Bridges

The most commonly used countermeasures for bridge structures are features incorporated into the design of the structure to reduce the potential for trapping and accumulating debris. Unfortunately, specific guidance or guidelines do not presently exist for these design features. However, general guidance is presented below.

1. FREEBOARD. Freeboard is a safety precaution of providing additional space between the design water surface elevation and the low chord elevation of the bridge. Considerations to the delivery potential of floating debris should be made in defining the amount of freeboard for a proposed bridge structure. When the potential for floating debris is remote or relatively low, freeboard is less important, whereas a careful selection of the freeboard is required for a bridge over a stream with a high potential for floating debris. The minimum freeboard of a bridge structure should be 0.6 meters (2.0 feet) where there is a high potential for floating debris. The freeboard should be increased to 1 to 1.2 meters (3.3 to 3.9 ft) where debris

is abundant and known debris problems exist. Unfortunately, freeboard alone cannot guarantee the complete elimination of damage because the degree of protection is limited by the ever-present chance that a flood will occur that exceeds the level of protection provided by the freeboard.

Increasing freeboard will decrease the probability of debris hazards to a certain degree; however, the cost of construction may increase significantly depending on the geometry of the river crossing and the bridge. For such locations, a cost-risk analysis should be performed to establish the recommended freeboard at the site.

2. PIER TYPE, LOCATION AND SPACING. As indicated in Chapters 2 and 3 of this manual, the potential for debris accumulation at a bridge structure is significantly influenced by the pier type, location and spacing. Therefore, these features should be evaluated during the design of proposed and replacement bridges where there is a high potential for debris delivery to the site. Piers located within the path of floating debris can have a high potential for accumulation even if the span length between the piers is significantly greater than the maximum length of the floating debris; or, piers that have adequate spacing and are out of the debris path can still have a high potential to accumulate debris if the piers have narrow openings that can easily trap debris.

The type of pier can influence the potential for debris to become trapped rather than deflected. Piers with narrow openings that convey flow are significantly more likely to trap and accumulate debris than piers without openings. Therefore to minimize the potential for entrapment, the bridge piers should be solid, round-nosed piers that are aligned with the approaching flow. If multiple columns are used, then considerations should be made to reduce the potential entrapment of debris between the columns by providing a solid web wall between the columns.

As previously stated, debris accumulations exist most frequently and in the greatest amount where the path of floating drift encounters fixed objects that divide the flow. Therefore, bridge piers should be placed outside of the debris path, which can be estimated using the information provided in Chapter 3 of this manual. In general, for a curved channel reach, piers should not be located near the bank toe on the outside bend, and in a straight reach, piers should not be located near the thalweg of the channel where the flow is the deepest and fastest. For critical locations, the piers should not be placed within the main channel, if this can be avoided.

The span length can influence the type of debris accumulation occurring at the bridge and the overall width of the accumulation. If the span length is less than the design log length, debris could become lodged between two piers or between a pier and the adjacent abutment and potentially block the entire span opening, i.e., span-blockage accumulation. Debris for this type of accumulations can extend beyond the piers, so the total width of the accumulation could be greater than the design log length. On the other hand if the span length is greater than the design log length, debris only accumulates on the piers at a width approximately equal to the design log length, i.e., single-pier accumulations. As a minimum, the span length should be slightly greater than the design log length, which can be determined using the recommendations by Diehl[17] provided on page 3.6 of this manual. Pier spacing should be even greater for streams with a high potential for debris delivery to the site since longer spans are less prone to debris blockage. Since the total cost of the bridge generally rises with increasing pier spacing and span length, the total cost of the bridge in relation to the pier spacing should be carefully evaluated.

3. SUPERSTRUCTURE. Where debris hazards persist and there is a high chance of the bridge being overtopped, the design of the superstructure should take into account the consequence of overtopping. The superstructure should be designed to withstand extreme floods even in a submerged condition. A thin deck and low railings could be incorporated into the design of the superstructure to minimize the lateral hydraulic forces on the structure. Also, the superstructure should be designed to minimize the potential for debris accumulation on the structure by eliminating any unnecessary narrow openings in the structure, i.e., solid parapet walls in lieu of open railings, and at the connection with the pier, i.e., a solid beam that is connected directly to the pier that would entrap and accumulate debris.

6.4   NON-STRUCTURAL GUIDELINES (DEBRIS MANAGEMENT)

The implementation of a debris management plan might be a cost effective method for structures on small watersheds. The purpose of this plan is to reduce excessive debris input into the stream network by clearing trash, debris jams and downed trees from the channel and floodplain of a stream and/or through multipurpose channel stabilization schemes. Large woody debris within a channel is a beneficial and vital geomorphologic and ecologic component of a river system [59,60,61] and the plan should recognize these benefits. Wallerstein and Thorne have developed such a plan[61] by taking into account the relationship between the large woody debris jam formations and channel processes discussed in Section 2.2 of this manual. This plan is summarized in Table 6.2.

Based on information of a given reach, the management plan provides information on the type of debris jam formation most likely to be present within the reach, impacts on the channel morphology associated with the type of debris jam formation, and an appropriate management strategy for the reach. The information required for the given reach includes the vegetation type, average riparian tree height, average channel width, and the type of sediment within the reach. The vegetation type is defined as either forest, agricultural, or open water with forest being the only type where substantial jams can form. The ratio of tree height to channel width is used to define the type of debris jams most likely to be present within the reach. The precise limits used to define the type of debris jams were determined from empirical relationships developed from field studies. The type of sediment, either fine (sand) or coarse (gravel) detritus, is used to distinguish if backwater sediment wedges or downstream bars will occur at the jams.

Table 6.2. Management Plan for the Large Woody Debris Formations.

Vegetation Type	Vegetation Height/ Channel Width	Sediment	Management Strategy[61]
Agricultural or Open Space	n.a.	n.a.	Substantial debris jams are unlikely to form within the reach since the immediate riparian zone is agricultural land or open water. Therefore, debris removal is unnecessary. Artificial debris input may be desirable for habitat enhancement, stabilization of sand bed channels through backwater sediment retention, or to reduce bank velocities on the outside of meanders.
Forest	1.3W ≤ H	n.a.	UNDERFLOW jams exist within the reach. Debris clearance is unnecessary since there would be minimal adverse geomorphic impacts associated with the jams (local scour may occur under the debris at high flows) and a significant quantity of heavy floating debris would unlikely be transported downstream. Therefore, bridge and other structures in the reach should not be affected by persistent debris accumulations.
Forest	0.95W < H < 1.3W	Coarse Detritus	DAM jams exist within the reach. Jams may cause significant local bed scour and bank erosion due to flow constriction. Backwater sediment wedges and bars may form upstream of the jams since the sediment consist of coarse detritus. The jams may also increase the duration of overbank flooding. A limited amount of floating debris may be transported downstream from the reach. Debris clearance may be necessary if the local bed and bank scour results in a significant increase of large woody debris being introduced into the stream.
Forest	0.95W < H < 1.3W	Fine Detritus	DAM jams exist within the reach. Jams may cause significant local bed scour and bank erosion due to flow constriction. Backwater sediment wedges and bars are unlikely to form upstream of the jams since the sediment consist of fine detritus. The jams may also increase the duration of overbank flooding. A limited amount of floating debris may be transported downstream from the reach. Debris clearance may be necessary if the local bed and bank scour results in a significant increase of large woody debris being introduced into the stream.
Forest	0.60W ≤ H ≤ 0.95W	Coarse Detritus	DEFLECTOR jams exist within the reach. Jams may cause significant bank erosion of one or both banks that could result in a significant increase of large woody debris being introduced into the stream. Since the sediment consist of coarse detritus, local bed scour induced by the jams will most likely be negligible and backwater sediment wedges may form upstream of the jams. Debris clearance unnecessary except where localized bank erosion results in a significant increase of large woody debris being introduced into the stream.
Forest	0.60W ≤ H ≤ 0.95W	Fine Detritus	DEFLECTOR jams exist within the reach. Jams may cause significant bank erosion of one or both banks that could result in a significant increase of large woody debris being introduced into the stream. Since the sediment consist of fine detritus, local bed scour induced by the jams might be significant and backwater sediment wedges and bars would most likely not form upstream of the jams. Debris clearance necessary to prevent local bank erosion.
Forest	H < 0.60W	n.a.	FLOW PARALLEL jams exist within the reach. Large woody debris will be transported downstream in high flows from this reach and deposited at bank base in meanders and at run-of-river structures. Adverse geomorphic impacts associated with the jams are minimal. Banks may be stabilized due to debris build-up, and debris may also accelerate formation of mid-channel bars. Debris clearance from channel unnecessary if it is keyed into place at bank toes and bars.

CHAPTER 7 – MAINTENANCE

Maintenance is a vital component in how a debris-control countermeasure would function at a drainage structure. The lack of maintenance can cause improper functioning of the structure, resulting in possible damage to or failure of the structure. Although no specific guidelines for highway maintenance practices have been established, the general maintenance practice of these structures should involve regular inspections and cleaning, coupled with emergency removal of debris. Regular periodic inspections allow minor problems to be discovered and corrected before they become serious. The procedure and documentation of the inspection should be based on the guidelines provided in FHWA's Culvert Inspection Manual[4] and FHWA's Bridge Inspector's Training Manual[27]. Debris accumulated on these structures should be removed in a timely manner since the presence of debris can have several severe consequences as discussed in Chapter 2 of this manual. In addition to the adverse effects associated with debris accumulations, structures with an existing debris accumulation have a higher potential for trapping additional debris than if they were cleared of debris. Emergency debris removal during a flood event can be critical to the survival of a structure laden with debris and to the flooding impacts upstream of the structure.

The frequency of maintenance must be considered in the design of these structures. Structures located on a primary highway may have a higher frequency of maintenance than those on a secondary highway. If a low standard of maintenance is to be provided at a debris-control countermeasure or structure, it may be desirable to use a debris-control countermeasure that requires less attention, a combination of debris-control countermeasures, and/or a larger structure. This consideration may also determine the choice of alternatives when two or more debris-control countermeasures are being considered for a site.

As mentioned in Section 6.4 of this manual, provisions should be made for maintenance access to the debris-control countermeasure or culvert/bridge structure. Unfortunately, access is often difficult to provide for a debris-control countermeasure and structure, particularly when it is near an existing high embankment. Such installations, however, usually require less maintenance because of the additional storage available for the debris. If access roads are impractical and the risk associated with flooding is high, it may be necessary to provide an area near the structure that mechanical equipment, such as a crane, could be located for removing debris from the structure without disrupting highway traffic. Access should be provided to the substructure of bridges to ensure prompt and complete removal of debris accumulations on the structure. Tracked vehicles can usually be used to remove the debris during low flows at bridges over small streams. For bridges over large rivers, a barge might be required to remove the debris, so a launching site for the barge may be necessary at such a site.

Debris removed from a culvert or bridge should be disposed in an acceptable fashion. It should not be disposed directly downstream of the structure, ignoring the consequences to any structure further downstream. Debris should not be placed where it may be carried away by the stream flow or where it may block the drainage of an area. Potential disposal options include burning, burial, using it as firewood, using it as chipped wood for horticultural purposes if low grade wood, or using it for structural purposes if high grade wood. The latter two options are preferable as they may create some financial return for the operator, while burial is costly and burning is environmentally undesirable, especially if the wood has become contaminated in the river by toxins, hydrocarbons, or heavy metals. In addition, this process seems to have limited success in reducing the amount of debris.

Ideally, a maintenance plan that clearly defines the maintenance activities and the responsibility of the maintenance crew should be developed for a structure susceptible to debris problems. As a minimum, this plan should contain information about the various items discussed above, i.e., inspections, regular and emergency debris maintenance activities, frequency of maintenance, access, and disposal of the debris. In addition, the general location and maximum extent of debris removal as estimated using the procedures and recommendations provided in Chapter 3 of this manual can be provided for bridges.

Maintenance associated with debris removal should be considered in the design of a new or replacement bridge. As discussed in Section 6.4 of this manual, there are certain features that can be incorporated into the design to simplify debris removal. Solid wall piers that extend a short distance upstream of the bridge deck edge are easier to remove debris from and less likely to cause damage to the bridge structure during the removal operations than other pier types. Superstructures that allow access to the pier nose from directly above also ease debris removal, such as, a wide deck with a simple parapet and adequate load-bearing capacity for heavy equipment.

CHAPTER 8 – REFERENCES

1. Abt S.R., Brisbane, T.E., Frick, D.M. & McKnight, C.A., 1992, "Trash Rack Blockage in Supercritical Flow", Journal of Hydraulic Engineering, ASCE, December pp 1692-1696.

2. Apelt, C.J., 1986, "Flow Loads on Submerged Bridges: Institution of Engineers Australia (Queensland Division)", Queensland Division Technical Papers, v. 27, no. 19, Publication No. QBG 1756, p. 17-23.

3. Arcement, G.J., Jr., and Schneider, V.R., 1989, "Guide for Selecting Manning's Roughness Coefficients for Natural Channels and Flood Plain, USGS Water-Supply Paper 2339", Department of Interior, U.S. Geological Survey, Denver, CO.

4. Arnoult, J.D., 1986, "Culvert Inspection Manual, Supplement to the Bridge Inspector's Training Manual", FHWA-IP-86-2, Research and Development, Turner-Fairbank Highway Research Center, McLean, Virginia: U.S. Department of Transportation, Federal Highway Administration.

5. AASHTO, 2000, "LRFD Bridge Design Specifications", 2nd edition; section 3.18

6. Ballinger, C.A. and Drake, P.G., 1995, "Culvert Repair Practices Manual, Volume I", FHWA-RD-94-096, Research and Development, Turner-Fairbank Highway Research Center, McLean, Virginia: U.S. Department of Transportation, Federal Highway Administration.

7. Ballinger, C.A. and Drake, P.G., 1995, "Culvert Repair Practices Manual, Volume II", FHWA-RD-95-089, Research and Development, Turner-Fairbank Highway Research Center, McLean, Virginia: U.S. Department of Transportation, Federal Highway Administration.

8. Barnes, H.H., Jr., 1967, "Roughness Characteristics of Natural Channels," U.S. Geological Survey, Water Supply Paper 1849, Washington D.C.

9. Bates, D.W., Murphey, E.W., and Beam, M.G., 1971, "Traveling Screen for Removal of Debris from Rivers", U.S. Department of Commerce, National Oceanic and Atmospheric Administration, National Marine Fisheries Service.

10. Bradley, J.N., 1978, "Hydraulic of Bridge Waterways, Hydraulic Design Series No. 1", Federal Highway Administration, U.S. Department of Transportation, Second Edition, revised March 1978, Washington D.C.

11. Brice, J.C., 1975, "Air Photo Interpretation of the Form and Behavior of Alluvial Rivers", Final report to the U.S. Army Research Office – Durham, Washington University, St. Louis.

12. Brice, J.C., Blodgett, J.C., Carpenter, P.J., Cook, M.F., Craig, G.S., Jr., Eckhardt, D.A., Hines, M.S., Lindskov, K.L., Moore, D.O., Parker, R.S., Scott, A.G., and Wilson, K.V., 1978, "Countermeasures for Hydraulic Problems at Bridge, Volume 1 – Analysis and Assessment", FHWA-RD-78-162, U.S. Department of Transportation, Federal Highway Administration, Washington, D.C.

13. Brice, J.C., Blodgett, J.C., Carpenter, P.J., Cook, M.F., Craig, G.S., Jr., Eckhardt, D.A., Hines, M.S., Lindskov, K.L., Moore, D.O., Parker, R.S., Scott, A.G., and Wilson, K.V., 1978, "Countermeasures for Hydraulic Problems at Bridge, Volume 2 – Case Histories for Sites", FHWA-RD-78-162, U.S. Department of Transportation, Federal Highway Administration, Washington, D.C.

14. Chang, F.F.M., and Shen, H.W., 1979, "Debris Problems in the River Environment", FHWA-RD-79-62, U.S. Department of Transportation, Federal Highway Administration, Washington, D.C.

15. Chow, V.T., 1959, "Open Channel Hydraulics", McGraw-Hill Book Company.

16. Denson, K.H., 1982, "Steady-State Drag, Lift and Rolling Moment Coefficients for Inundated Inland Bridges", Mississippi Water Resources Research Institute, Mississippi State University.

17. Diehl, T.H., 1997, "Potential Drift Accumulation at Bridges", FHWA-RD-97-28, U.S. Department of Transportation, Federal Highway Administration, Washington, D.C.

18. Diehl, T.H., and Bryan, B.A., 1993, "Supply of Large Woody Debris in a Stream Channel", in Shen, H.W., Su, S.T., and Wen, Feng, eds., Hydraulic Engineering 1993 Conference, San Francisco, 1993, Proceedings: American Society of Civil Engineers, v.1, p. 1055-1060.

19. Doheny E.J., 1993, "Relation of Channel Stability to Scour at Highway Bridges over Waterways in Maryland", Hydraulic Engineering 1993, ASCE, pp 2243-2248.

20. Dongol, M.S., 1989, "Effect of Debris Rafting on Local Scour at Bridge Piers", Report No. 473, School of Engineering, University of Auckland, Auckland, New Zealand.

21. Fiebiger, G., 1993, "Structures of Debris Flow Countermeasures", Debris-Flow Hazards Mitigation: Mechanics, Prediction, and Assessment, pp 596-605.

22. Fox, S.V., Gilje, S.A., Johnson, F.L., Kober, W., McGuffey, V.C., Reed, L., Robinson, A.R., Spaine, L.F., and Zanganeh, S., 1980, "Design of Sedimentation Basins", National Cooperative Highway Research Program Synthesis of Highway Practice 70, Transportation Research Board, Washington, D.C.

23.   Froehlich, H.A., 1973, "Natural and Man-Caused Slash in Headwater Streams", Pacific Logging Congress, Loggers Handbook, Vol. 33, Oregon State University, School of Forestry.

24.   Genetti, A.J. Jr., 1995, "EM 1110-2-4000, Sedimentation Investigation of Rivers and Reservoirs," U.S. Army Corps of Engineers, Washington D.C.

25.   Gilje, S.A., 1979, "Debris Problems in the River Environment", Public Roads.

26.   Haile, H.H., 1982, "Criteria for Design of Debris Dams and Basins", Los Angeles County Flood Control District.

27.   Hartle, R. A.; Amrhein, W. J.; Wilson, K. E.; Baughman, D. R.; and Tkacs, J. J., 1995, "Revised Bridge Inspector Training Manual", FHWA-PD-91-015, Research and Development, Turner-Fairbank Highway Research Center, McLean, Virginia: U.S. Department of Transportation, Federal Highway Administration.

28.   Hicks, D.M. and Mason, P.D., 1991, "Roughness Characteristics of New Zealand Rivers", Water Resources Survey, DSIR Marin and Freshwater, New Zealand.

29.   Hydrological Engineering Center, 1995, "Flow Transitions in Bridge Backwater Analysis, RD-42", U.S. Army Corps of Engineers, Davis, CA.

30.   Hydrological Engineering Center, 2001, "HEC-RAS, River Analysis System, Hydraulic Reference Manual", U.S. Army Corps of Engineers, Davis, CA.

31.   King, H.W. and Brater, E.F., 1963, "Handbook of Hydraulics", Fifth Edition, McGraw Hill Book Company, New York.

32.   Johnson, P.A., and McCuen, R.H., 1989, "Silt Dam Design for Debris Flow Mitigation", ASCE Journal of Hydraulic Engineering, Vol. 115, No. 9, pp. 1293 – 1296.

33.   Johnson, P.A., Hey, R.D., Horst, M.W., and Hess, A.J., 2001, "Aggradation at Bridges", ASCE Journal of Hydraulic Engineering, Vol. 127, No. 2, pp. 154 – 157.

34.   Lagasse, P.F., Schall, J.D., and Richardson, E.V., 2001, "Stream Stability at Highway Structures", Hydraulic Engineering Circular No. 20, Third Edition, FHWA NHI 01-002, Federal Highway Administration, Washington, D.C.

35.   Lagasse, P.F., Zevenbergen, L.W., Schall, J.D., and Clopper, P.E., 2001, "Bridge Scour and Stream Instability Countermeasures", Hydraulic Engineering Circular No. 23, Second Edition, FHWA NHI 01-003, Federal Highway Administration, Washington, D.C.

36.   Lagasse, P.F., Zevenbergen, L.W., Spitz, W.J., and Thorne, C.R., 2003, "Methodology for Predicting Channel Migration", NCHRP Project No. 24-16, Ayres Associations for

National Research Council, Transportation Research Board, National Cooperative Highway Research Program, Washington, D.C.

37.   Lagasse, P.F., Spitz, W.J., Zevenbergen, L.W., and Zachmann, D.W., 2003, "Handbook for Predicting Stream Meander Migration using Aerial Photographs and Maps", Ayres Associations for National Research Council, Transportation Research Board, National Cooperative Highway Research Program, Washington, D.C.

38.   MacArthur, R.C., Hamilton D., and Gee M.D., 1995, "Application of Methods and Models for Prediction of Land Surface Erosion and Yield", U.S. Army Corps of Engineers, Hydrologic Engineering Center, Davis, CA.

39.   Melville, B.W., and Dongol, D.M., 1992, "Bridge Pier Scour with Debris Accumulation", ASCE Journal of Hydraulic Engineering, Vol. 118, No. 9, pp. 1306 - 1310.

40.   McFadden, T., and Stallion, M., 1976, "Debris of the Chena River", CRREL Report 76-26, U.S. Army Corps of Engineers.

41.   Mueller, D.S., and Parola, A.C., 1998, "Detailed Scour Measurements Around a Debris Accumulation", Water Resource Engineering 1998, ASCE, pp. 234 – 239.

42.   Naudascher, E. and Medlarz, H.J., 1983, "Hydrodynamic Loading and Backwater Effect of Partially Submerged Bridges", Journal of Hydraulic Research, Vol. 21, No. 3, pp. 213 - 231.

43.   Ng Y.L.A., and Richardson, J.R., 2001, "Transport Mechanics of Floating Woody Debris", Thesis project presented to Faculty of the Graduate School, University of Missouri-Columbia.

44.   Odgaard, A.J. and Spoljaric A., 1986, "Sediment Control by Submerged Vanes", ASCE Journal of Hydraulic Engineering, Vol. 112, No. 12, pp 1164-1181.

45.   Odgaard, J.A., and Wang, Y., 1990, "Sediment Control in Bridge Waterway", IIHR Report No. 336, Iowa Institute of Hydraulic Research.

46.   Odgaard, A.J., and Wang, Y., 1991, "Sediment Management with Submerged Vanes II: Applications", ASCE Journal of Hydraulic Engineering, Vol. 117, No. 3, pp 284-302.

47.   O'Donnell, C.L., 1973, "Observations on the Causes of Bridge Damage in Pennsylvania and New York due to Hurricane Agnes", Highway Research Record, No. 479.

48.   Okubo, S., Ikeya, H., Ishikawa, Y., and Yamada, T., 1997, "Development of New Methods for Countermeasures against Debris Flow", Recent developments on debris flow, Debris avalanches – Congresses, pp. 166 – 185.

49. Parola, A.C., 2000, "Debris Forces on Highway Bridges", <u>NCHRP 445</u>, National Cooperative Highway Research Program.

50. Parola, A.C., Hagerty, D.J., and Kamojjala, S., 1998, "Highway Infrastructure Damage Caused by the 1993 Upper Mississippi River Basin Flooding", <u>NCHRP 12-39</u>, National Cooperative Highway Research Program.

51. Parola, A.C., and Kamojjala, S., Richardson, J., and Kirby, M., 1998, "Numerical Simulation of Flow Patterns at a Bridge with Debris", <u>Water Resource Engineering 1998,</u> pp. 240 - 245.

52. Perham, R.E, 1987, "Floating Debris Control: A Literature Review", <u>Technical Report No. REMR-HY-2</u>, U.S. Army Corps of Engineers.

53. Reihsen, G., and Harrison, L.J., 1971, "Debris-Control Structures", <u>Hydraulic Engineering Circular No. 9</u>, U.S. Department of Transportation, Federal Highway Administration, Washington, D.C.

54. Richardson, E.V. et at., 1975, "Highway in the River Environment, Hydraulics and Environmental Design Considerations, Training, and Design Manual", U.S. Department of Transportation, Federal Highway Administration, Washington, D.C.

55. Richardson, E.V., and Davis, S.R., 2001, "Evaluating Scour at Bridges", <u>Hydraulic Engineering Circular No. 18</u>, Fourth Edition, FHWA NHI01-001, Federal Highway Administration, Washington, D.C.

56. Saunders, S., and Oppenheimer, M.L., 1993, "A Method of Managing Floating Debris", <u>Hydraulic Engineering 1993</u>, ASCE, pp 1373-1378.

57. Savage, Jerry, 2003, Photograph of Oklahoma Bridge Debris failure, project correspondence

58. Vanoni, V.A., 1977, "Sedimentation Engineering", <u>ASCE Manual and Reports on Engineering Practice – No. 54</u>, ASCE.

59. Wallerstein, N.,1997, "Impact of Large Woody Debris on Fluvial Processes and Channel Geomorphology in Unstable Sand Bed River", Ph. D Thesis presented to the University of Nottingham.

60. Wallerstein, N. and Thorne, C.R., 1995, "Impacts of Wood Debris on Fluvial Processes and Channel Morphology in Stable and Unstable Streams", U.S. Army Research, Development and Standardization Group-UK.

61. Wallerstein, N. and Thorne, C.R., 1996, "Debris Control at Hydraulic Structures – Management of Woody Debris in Natural Channels and at Hydraulic Structures", U.S. Corps of Engineers, Waterways Experiment Station.

62.   Wallerstein, N. and Thorne, C.R., 1998, "Computer Model for Prediction of Scour at Bridges Affected by Large Woody Debris", International Water Resources Engineering Conference.

63.   Wallerstein, N., Thorne, C.R., Abt S.R., 1997, "Debris Control at Hydraulic Structures in Selected Areas of the United States and Europe", U.S. Army Engineer Waterways Experiment Station.

64.   Wellwood, N. and Fenwick, J., "A Flood Loading Methodology for Bridges," Conference Proceedings, 15th ARRB Conference, part 3, 1989, pp. 315-341.

65.   Williams, J.R., and Berndt, H.D., 1972, "Sediment Yield Computed with Universal Equation', ASCE Journal of Hydraulic Division, Vol. 98, No HY12, pp. 2087-2097.

66.   Yarnell, D.L., 1934, "Bridge Piers as Channel Obstructions," Technical Bulletin 442, U.S. Department of Agriculture, Washington D.C.

APPENDIX A – METRIC SYSTEM, CONVERSIONS, AND WATER PROPERTIES

(page intentionally left blank)

APPENDIX A

Metric System, Conversion Factors, and Water Properties

The following information is summarized from the Federal Highway Administration, National Highway Institute (NHI) Course No. 12301, "Metric (SI) Training for Highway Agencies." For additional information, refer to the Participant Notebook for NHI Course No. 12301.

In SI there are seven base units, many derived units and two supplemental units (Table A.1). Base units uniquely describe a property requiring measurement. One of the most common units in civil engineering is length, with a base unit of meters in SI. Decimal multiples of meter includes the kilometer (1000 m), the centimeter (1 m/100) and the millimeter (1 m/1000). The second base unit relevant to highway applications is the kilogram, a measure of mass which is the inertial of an object. There is a subtle difference between mass and weight. In SI, mass is a base unit, while weight is a derived quantity related to mass and the acceleration of gravity, sometimes referred to as the force of gravity. In SI the unit of mass is the kilogram and the unit of weight/force is the Newton. Table A.2 illustrates the relationship of mass and weight. The unit of time is the same in SI as in the English system (seconds). The measurement of temperature is Centigrade. The following equation converts Fahrenheit temperatures to Centigrade, $°C = 5/9$ $(°F - 32)$.

Derived units are formed by combining base units to express other characteristics. Common derived units in highway drainage engineering include area, volume, velocity, and density. Some derived units have special names (Table A.3).

Table A.4 provides useful conversion factors from English to SI units. The abbreviations presented in this table for metric units, including the use of upper and lower case (e.g., kilometer is "km" and a Newton is "N") are the standards that should be followed. Table A.5 provides the standard SI prefixes and their definitions.

Tables A.6 and A.7 provide physical properties of water at atmospheric pressure in SI and English systems of units, respectively. Table A.8 gives the sediment grade scale and Table A.9 gives some common equivalent hydraulic units.

Table A.1. Overview of SI.		
	Units	Symbol
Base units		
length	meter	m
mass	kilogram	kg
time	second	s
temperature*	kelvin	K
electrical current	ampere	A
luminous intensity	candela	cd
amount of material	mole	mol
Derived units	**	
Supplementary units angles in the plane solid angles	radian steradian	rad sr
* Use degrees Celsius (_C), which has a more common usage than kelvin. ** Many derived units exist; several will be discussed in this session.		

Table A.2 Relationship of Mass and Weight.			
	Mass	Weight or Force of Gravity	Force
English	slug pound-mass	pound pound-force	pound pound-force
metric	kilogram	newton	newton

Table A.3. Derived Units With Special Names.			
Quantity	Name	Symbol	Expression
Frequency	hertz	Hz	s^{-1}
Force	newton	N	$Kg \cdot m/s^2$
Pressure, stress	pascal	Pa	N/m^2
Energy, work, quantity of heat	joule	J	$N \cdot m$
Power, radiant flux	watt	W	J/s
Electric charge, quantity	coulomb	C	$A \cdot s$
Electric potential	volt	V	W/A
Capacitance	farad	F	C/V
Electric resistance	ohm	Ω	V/A
Electric conductance	siemens	S	A/V
Magnetic flux	weber	Wb	$V \cdot s$
Magnetic flux density	tesla	T	Wb/m^2
Inductance	henry	H	Wb/A
Luminous flux	lumen	lm	$cd \cdot sr$
Illuminance	lux	lx	lm/m^2

Table A.4. Useful Conversion Factors.			
Quantity	From English Units	To Metric Units	Multiplied by*
Length	Mile	km	1.609
	yard	m	0.9144
	foot	m	0.3048
	inch	mm	25.4
Area	square mile	km^2	2.590
	acre	m^2	4047
	acre	hectare	0.4047
	square yard	m^2	0.8361
	square foot	m^2	0.092 90
	square inch	mm^2	645.2
Volume	acre foot	m^3	1 233
	cubic yard	m^3	0.7646
	cubic foot	m^3	0.028 32
	cubic foot	L (1000 cm^3)	28.32
	100 board feet	m^3	0.2360
	gallon	L (1000 cm^3)	3.785
	cubic inch	cm^3	16.39
Mass	Lb	kg	0.4536
	kip (1000 lb)	metric ton (1000 kg)	0.4536
Mass/unit length	plf	kg/m	1.488
Mass/unit area	psf	kg/m^2	4.882
Mass density	pcf	kg/m^3	16.02
Force	lb	N	4.448
	kip	kN	4.448
Force/unit length	plf	N/m	14.59
	klf	kN/m	14.59
Pressure, stress, modulus of elasticity	psf	Pa	47.88
	ksf	kPa	47.88
	psi	kPa	6.895
	ksi	MPa	6.895
Bending moment, torque, moment of force	ft-lb	N · m	1.356
	ft-kip	kN · m	1.356
Moment of mass	lb · ft	kg · m	0.1383
Moment of inertia	lb · ft^2	kg · m^2	0.042 14
Second moment of area	In^4	mm^4	416 200
Section modulus	In^3	mm^3	16 390
Power	ton (refrig)	kW	3.517
	Btu/s	kW	1.054
	hp (electric)	W	745.7
	Btu/h	W	0.2931

Table A.4. Useful Conversion Factors (continued).			
Quantity	From English Units	To Metric Units	Multiplied by*
Volume rate of flow	ft^3/s cfm cfm mgd	m^3/s m^3/s L/s m^3/s	0.028 32 0.000 471 9 0.4719 0.0438
Velocity, speed	ft/s	m/s	<u>0.3048</u>
Acceleration	f/s^2	m/s^2	<u>0.3408</u>
Momentum	lb · ft/sec	kg · m/s	0.1383
Angular momentum	lb · ft^2/s	kg · m^2/s	0.042 14
Plane angle	degree	rad mrad	0.017 45 17.45
* 4 significant figures; underline denotes exact conversion			

Table A.5. Prefixes					
Submultiples			Multiples		
deci	10^{-1}	d	deka	10^1	da
centi	10^{-2}	c	hector	10^2	h
milli	10^{-3}	m	kilo	10^3	k
micro	10^{-6}	μ	mega	10^6	M
nano	10^{-9}	n	giga	10^9	G
pica	10^{-12}	p	tera	10^{12}	T
femto	10^{-15}	f	peta	10^{15}	P
atto	10^{-18}	a	exa	10^{18}	E
zepto	10^{-21}	z	zeta	10^{21}	Z
yocto	10^{-24}	y	yotto	10^{24}	Y

Table A.6. Physical Properties of Water at Atmospheric Pressure in SI Units

Temperature		Density	Specific Weight	Dynamic Viscosity	Kinematic Viscosity	Vapor Pressure	Surface Tension[1]	Bulk Modulus
Centigrade	Fahrenheit	kg/m^3	N/m^3	N $\cdot$ s/m^2	^{2}/s	N/m^2 abs.	N/m	GN/m^2
0°	32°	1,000	9,810	1.79×10^{-3}	1.79×10^{-6}	611	0.0756	1.99
5°	41°	1,000	9,810	1.51×10^{-3}m	1.51×10^{-6}	872	0.0749	2.05
10°	50°	1,000	9,810	1.31×10^{-3}	1.31×10^{-6}	1,230	0.0742	2.11
15°	59°	999	9,800	1.14×10^{-3}	1.14×10^{-6}	1,700	0.0735	2.16
20°	68°	998	9,790	1.00×10^{-3}	1.00×10^{-6}	2,340	0.0728	2.20
25°	77°	997	9,781	8.91×10^{-4}	8.94×10^{-7}	3,170,	0.0720	2.23
30°	86°	996	9,771	7.97×10^{-4}	8.00×10^{-7}	4,250	0.0712	2.25
35°	95°	994	9,751	7.20×10^{-4}	7.24×10^{-7}	5,630	0.0704	2.27
40°	104°	992	9,732	6.53×10^{-4}	6.58×10^{-7}	7,380	0.0696	2.28
50°	122°	988	9,693	5.47×10^{-4}	5.53×10^{-7}	12,300	0.0679	
60°	140°	983	9,643	4.66×10^{-4}	4.74×10^{-7}	20,000	0.0662	
70°	158°	978	9,594	4.04×10^{-4}	4.13×10^{-7}	31,200	0.0644	
80°	176°	972	9,535	3.54×10^{-4}	3.64×10^{-7}	47,400	0.0626	
90°	194°	965	9,467	3.15×10^{-4}	3.26×10^{-7}	70,100	0.0607	
100°	212°	958	9,398	2.82×10^{-4}	2.94×10^{-7}	101,300	0.0589	

[1]Surface tension of water in contact with air

A.8

Table A.7. Physical Properties of Water at Atmospheric Pressure in SI Units

Temperature		Density	Specific Weight	Dynamic Viscosity	Kinematic Viscosity	Vapor Pressure	Surface Tension[1]	Bulk Modulus
Fahrenheit	Centigrade	$Slug/ft^3$	lb/ft^3	$lb\text{-}sec/ft^2$	ft^2/sec	lb/in^2	lb/ft	lb/in^2
32°	0°	1.940	62,416	0.374×10^{-4}	1.93×10^{-5}	0.09	0.00518	1.99
39.2	4.0°	1.940	62,424	0				
40°	4.4°	1.940	62,423	0.323	1.67	0.12	0.00514	2.05
50°	10.0°	1.940	62,408	0.273	1.41	0.18	0.00508	2.11
60°	15.6°	1.939	62,366	0.235	1.21	0.26	0.00504	2.16
70°	21.1°	1.936	62,300	0.205	1.06	0.36	0.00497	2.20
80°	26.7°	1.934	62,217	0.180	0.929	0.51	0.00492	2.23
90°	32.2°	1.931	62,118	0.160	0.828	0.70	0.00486	2.25
100°	37.8°	1.927	61,998	0.143	0.741	0.95	0.00479	2.27
120°	48.9°	1.918	61,719	0.117	0.610	1.69	0.0466	2.28
140°	60°	1.908	61,386	0.0979	0.513	2.89		
160°	71.1°	1.896	61,006	0.0835	0.440.	4.74		
180°	82.2°	1.883	60,586	0.0726	0.385	7.51		
200°	93.3°	1.869	60,135	0.0637	0.341	11.52		
212°	100°	1.847	59,843	0.0593	0.319	14.70		

[1]Surface tension of water in contact with air, weight of sea water approximately 63.93 lb/ft^3 @ 15°C

Table A.8. Sediment Particles Grade Scale.

Size				Approximate Sieve Mesh Opening per Inch		Class
Millimeters	Millimeters	Microns	Inches	Tyler	U.S. Standard	
4000-2000	----	----	160-80	----	----	Very large boulders
2000-1000	----	----	80-40	----	----	Large boulders
1000-500	----	----	40-20	----	----	Medium boulders
500-250	----	----	20-10	----	----	Small boulders
250-130	----	----	10-5	----	----	Large cobbles
130-64	----	----	5-2.5	----	----	Small cobbles
64-32	----	----	2.5-1.3	----	----	Very coarse gravel
32-16	----	----	1.3-0.6	----	----	Coarse gravel
16-8	----	----	0.6-0.3	2 1/2	----	Medium gravel
8-4	----	----	0.3-0.16	5	5	Fine gravel
4-2	----	----	0.16-0.08	9	10	Very fine gravel
2-1	2.00-1.00	2000-1000	----	16	18	Very coarse sand
1-1/2	1.00-0.50	1000-500	----	32	35	Coarse sand
1/2-1/4	0.50-0.25	500-250	----	60	60	Medium sand
1/4-1/8	0.25-0.125	250-125	----	115	120	Fine sand
1/8-1/16	0.125-0.062	125-62	----	250	230	Very fine sand
1/16-1/32	0.062-0.031	62-31	----	----	----	Coarse silt
1/32-1/64	0.031-0.016	31-16	----	----	----	Medium silt
1/64-1/128	0.016-0.008	16-8	----	----	----	Fine silt
1/128-1/256	0.008-0.004	8-4	----	----	----	Very fine silt
1/256-1/512	0.004-.0020	4-2	----	----	----	Coarse clay
1/512-1/1024	0.0020-0.0010	2-1	----	----	----	Medium clay
1/1024-1/2048	0.0010-0.0005	1-0.5	----	----	----	Fine clay
1/2048-1/4096	0.0005-0.0002	0.5-0.24	----	----	----	Very fine clay

Table A.9. Common Equivalent Hydraulic Units

Volume

Unit	Equivalent							
	cubic inch	liter	u.s. gallon	cubic foot	cubic yard	cubic meter	acre-foot	sec-foot-day
liter	61.02	1	0.264 2	0.035 31	0.001 308	0.001	810.6 E - 9	408.7 E - 9
u.s. gallon	231.0	3.785	1	0.1337	0.004 951	0.003 785	3.068 E - 6	1.547 E - 6
cubic foot	1728	28.32	7.481	1	0.037 04	0.028 32	22.96 E - 6	11.57 E - 6
cubic yard	46,660	764.6	202.0	27	1	0.746 6	619.8 E - 6	312.5 E - 6
meter^3	61,020	1000	264.2	35.31	1.308	1	810.6 E - 6	408.7 E - 6
acre-foot	75.27 E + 6	1,233,000	325,900	43 560	1.613	1 233	1	0.504 2
sec-foot-day	149.3 E + 6	2,447,000	646,400	66 400	3 200	2 447	1.983	1

Discharge (Flow Rate, Volume/Time)

Unit	Equivalent					
	gallon/min	liter/sec	acre-foot/day	foot^3/sec	million gal/day	meter^3/sec
gallon/minute	1	0.063 09	0.004 419	0.002 228	0.001 440	63.09 E - 6
liter/second	15.85	1	0.070 05	0.035 31	0.022 82	0.001
acre-foot/day	226.3	14.28	1	0.504 2	325 9	0.014 28
feet^3/second	448.8	28.32	1.983	1	0.646 3	0.028 32
meter^3/second	15,850	1000	70.04	35.31	22.83	1

(page intentionally left blank)

APPENDIX B – STATE SURVEY

(page intentionally left blank)

APPENDIX B – STATE SURVEY

State	Information
Texas	Provided a web site for their hydraulic manual. This manual generally flags the issue and provides some general guidelines (use a single box in lieu of a double box, multiple boxes at different elevation, increase freeboard, and increase span lengths). It is indicated in the response that problems associated with debris doesn't appear to be a major problem for the state.
Oklahoma	Devices to control debris: debris sweeper, increase freeboard, and single or triple cell RCB in lieu of double RCB. No information available for debris sweeper since it is a new installation.
Michigan	No design procedures, practice, devices or strategies have been employed to control debris accumulation problem; other than, immediately remove any debris accumulated at the structures.
Missouri	No standard plans for debris control structures, and they are not routine used in their drainage designs.
Connecticut	Discourage the use of debris control structures on culverts. Use rounded nose for multiple culvert or for bridge piers. Debris racks have been used with varying degrees of success. However, these structures are discouraged except where absolutely necessary because they are invariably maintenance intensive.
Virginia	Very limited experience with debris control structures. A few structures were constructed about 25 to 30 years ago. Some or possibly all of these structures have been damaged in storms and have been removed.
Montana	Large culverts – They have tried H-piles placed upstream of the culvert on a limited basis. Bridges – location of piers, minimize number of piers, maintain adequate freeboard, and removal of any debris accumulations (maintenance).
South Dakota	Multiple barrel box culverts – extend the upstream end of the interior walls to the end of the apron with a height at the upstream end set at or above a computed water surface elevation (sloping nose). Bridges – use pier walls instead of series of columns. Safety bars at the upstream end of the culverts have services as debris racks and have reduced the culvert performance.
Mississippi	They have used drift deflectors and web walls for bridges.
Kansas	Bridges – no piers in the main channel, webwall for the width of the pier, align the piers to the streamflow, structure sized assuming a debris raft is present (drift potential is determined based on historical records, photos, and a site visit). Structures that span the main channel show limited amount of drift buildup.
Florida	District 3 will be installing a debris sweeper on one of their bridges, and they will be using this device on several other bridges depending on how it functions on the first bridge. Some of the local bridge owners have installed sacrificial 18" piling immediately upstream of the bridge pier to essentially shift the debris buildup and cleanup away from the structure.

State	Information
Indiana	Recommend 80' to 90' minimum span lengths for large stream crossings, and they have used debris deflectors. They are currently performing a research project to assess the success rate of the pile debris deflectors.
South Carolina	They have designed various structures on a case by case basis using the old version of HEC-9 as a guideline. They have no standard plans for these structures. There have been no significant complaints made by their field office.
Arkansas	They have not used any debris control structure at any culverts or bridges.
Tennessee	Have used debris fins at the inlet of culverts. They have also installed debris sweeper at two bridge sites.
Kentucky	They have not used any debris control structure at any culverts or bridges.

Revisions and Errata

November 16, 2005 – Joe Krolak – Final version of second edition & pdf

October 31, 2007 – Joe Krolak – Corrected Debris Rack height and location (subsection 6.2.2) based on Eric Brown and USFS inquiry of January 2007.

www.ingramcontent.com/pod-product-compliance
Lightning Source LLC
Chambersburg PA
CBHW080638180526

45168CB00008B/3217